Privacy, Surveillance, and the New Media You

Steve Jones
General Editor

Vol. 96

The Digital Formations series is part of the Peter Lang Media and Communication list.
Every volume is peer reviewed and meets
the highest quality standards for content and production.

PETER LANG
New York • Bern • Frankfurt • Berlin
Brussels • Vienna • Oxford • Warsaw

Edward Lee Lamoureux

Privacy, Surveillance, and the New Media You

HUMBER LIBRARIES LAKESHORE CAMPUS
3199 Lakeshore Blvd West
TORONTO, ON.   M8V 1K8

PETER LANG
New York • Bern • Frankfurt • Berlin
Brussels • Vienna • Oxford • Warsaw

Library of Congress Cataloging-in-Publication Data
Names: Lamoureux, Edward Lee, author.
Title: Privacy, surveillance, and the new media you / Edward Lee Lamoureux.
Description: New York: Peter Lang, 2016.
Series: Digital formations, vol. 96 | ISSN 1526-3169
Includes bibliographical references and index.
Identifiers: LCCN 2016030615 | ISBN 978-1-4331-2495-2 (hardcover: alk. paper)
ISBN 978-1-4331-2494-5 (paperback: alk. paper) | ISBN 978-1-4539-1824-1 (ebook pdf)
ISBN 978-1-4331-3579-8 (epub) | ISBN 978-1-4331-3580-4 (mobi)
Subjects: LCSH: Privacy, Right of—United States.
Electronic surveillance—United States.
Classification: LCC JC596.2.U6 L35 2016 | DDC 323.44/820973—dc23
DOI: 10.3726/b10478
LC record available at https://lccn.loc.gov/2016030615

Bibliographic information published by **Die Deutsche Nationalbibliothek**.
Die Deutsche Nationalbibliothek lists this publication in the "Deutsche
Nationalbibliografie"; detailed bibliographic data are available
on the Internet at http://dnb.d-nb.de/.

The paper in this book meets the guidelines for permanence and durability
of the Committee on Production Guidelines for Book Longevity
of the Council of Library Resources.

© 2016 Edward Lee Lamoureux
Peter Lang Publishing, Inc., New York
29 Broadway, 18th floor, New York, NY 10006
www.peterlang.com

All rights reserved.
Reprint or reproduction, even partially, in all forms such as microfilm,
xerography, microfiche, microcard, and offset strictly prohibited.

Printed in the United States of America

CONTENTS

Acknowledgements	ix
Preface	xi
Introduction	xvii
How the Heck Did We Get Here?	xxiii
Technological Complexities in Data Collection: Out of Sight, Out of Reach	xxvii
Appealing to Consumers' Lowest Common Denominators: It's All Good?	xxxii
Supporting Economic Interests of Business and Industry: Doubling Down on Data	xxxiii
Unwillingness of Government to Protect Its Citizenry from Itself: Corporations Are Our Most Valued People	xxxvi
Concerns over Terrorism and Security Risks: A Darned (In)Convenient War on Terror	xxxvii
Family Observations: A Personal Anecdote	xxxvii
Chapter Previews	xxxix

Chapter 1: The Harms	1
Loss of Control over Personal Data	2
Hacking as Lawlessness	2
Leaks and Sponges: We Were Not Supposed to Collect It or Lose It	4
Costs and Harms of Data Breaches	7
Undermining Constitutional Protections	8
Encroaching on First Amendment Protections	8
Encroaching on Third Amendment Protections	13
Encroaching on Fourth Amendment Protections	15
Encroaching on Fifth Amendment Protections	17
Encroaching on Fourteenth Amendment Protections	18
Article III: No Harm No Foul? Perhaps Not So Much	22
Personal Data and the Value Equation	23
Chapter 2: Centerings	29
Theoretical Orientations	29
Interrogating New Media	32
The Framework of Contextual Integrity	34
Identifying Loci of Control	36
Ideological Powers in Platforms	38
Chapter 3: FIP 1: No Secret Data Collection	41
When Is a Contract Just Barely, or Not, a Contract?	43
Data Anonymity via Aggregation as Oxymoron	49
Even the Most Private Datum Isn't	51
Knowing about the Unknowable	55
Quick Reminder: The Constitution Is Supposed to Protect Our Privacy	57
Chapter 4: FIPs 2 and 4: Discovery and Repair	65
You Can't Fix It If You Don't Know About—And Can't Access—It	67
Online Privacy and the European Union	74
The Right to Be Forgotten	76
Ownership of Social Media Accounts in the Enterprise: "Unhand My Stuff!"	77
Toward Chapter 5 and the Third FIP	79
Chapter 5: FIP 3: One Use Should Not Bind Them All	81
Data Collection and Security: Commercial Entities Sharing with the Government	86

Battles over Encryption and Cryptography	86
Encryption and Export Controls	93
The Snowden Revelations	96
Data Collection and Insecurity: The Government Sharing with Commercial Entities	99
Crossing Government and Commercial Streams within the Data Marketplace	104
We Claim to Know Better but Do Not Act Like It	109
Chapter 6: FIP 5: If You Don't Protect It, You Should Not Have It	113
Improved Data Reliability via the Marketplace	114
Big Data Can Aid Problem Solving	114
Recommendations Can Help	118
Improved Targeting Can Be Good for Both Sides	121
The Usability and Functionality Lies	124
Unreliability via the Marketplace	129
Data Abuse via the Marketplace	133
You Can Touch This Because I Can't Protect It	137
Chapter 7: Recommendations	141
Action Proposals for Government	146
Executive Leadership	146
Federal Agencies	150
Congressional Legislation	154
Congressional Legislative Action on Commercial Activities	154
Congressional Legislative Action on Intelligence Activities	158
Intelligence Community and Law Enforcement	159
Judges and Courts	163
State-Level Actions	167
States and the FIPs: Constraint of Government Agencies and Law Enforcement from Predatory Practices	168
State-Based Actions in the Commercial Privacy Environment	171
Action Proposals for Commercial Entities	173
Industry Leadership	173
Executive Leadership	176
Worker Behaviors	179

    Protocol Changes   181
        Privacy Policies and Terms of Service   181
        Flip the Defaults   182
        Offer Multiple Versions   183
        Reconsider Data as Revenue Streams   184
        Reengineer Data Anonymity   184
    Nonprofits: Great Responsibilities   185
WE, the People   191

Notes   203
Works Cited   205
About the Author   231
About the Front Cover   233
Index   235

ACKNOWLEDGEMENTS

Academic writers are enabled by their editors and publishers. In my case, good fortune has granted me wonderful working relationships with Peter Lang's Mary Savigar and series editor Steve Jones. Both are friends and valued professional colleagues. I've been able to, mostly, be good to my promises (if not our deadlines). Lang provides the best copyeditors and proofreaders in the business; my thanks to Alison Jacques and Suzie Tibor.

Former student, Bradley alum, and longtime friend Matt Crain provided both motivation and wisdom, by writing a fantastic doctoral dissertation that helped me better understand aspects of the topic that had otherwise alluded me. Additionally, Matt and I exchange roles as reader/writer—critical work that benefits us both. Bradley University's security guru, David Scuffham, spent time listening to my explanations and then guided me toward better understandings. BU senior system administrator Paul Carpenter has long been more patient than any geek should be with a dilettante such as me.

I am grateful for the support of my BU supervisors and colleagues, Ethan Ham, Paul Gullifor, and Jeffrey Huberman; my colleague Chris Kasch is always available when I need him. I am grateful to many undergraduate students who put up with the long-term development of the ideas in this book, across numerous offerings of not-quite-ready-for-prime-time privacy courses.

I toil in the shadows of my family: my long-suffering wife, Cheryl, and now-adult offspring Alexander, Samantha, Kate, and Nicole. They move me.

Any errors in the book are mine. Hopefully, the NSA, the data marketplace, and I can keep those closely guarded.

PREFACE

At the start, I will avoid using the word 'apocalypse' and the phrase 'the end of civilization as we know it.' However, I admit to membership in the Chicken Little and Paul Revere social clubs. The sky *is* falling and, although the British are not coming, we *are* under attack. People who research and write about privacy and surveillance run the risk of being labeled Luddites at best and perhaps fanatics or lunatics at worst. Even one, like myself, who attempts to address the subject in moderate and measured terms sometimes finds that the arguments they, or I, put forward appeal only to a segment of the audience that may well qualify for those pejoratives. I have given public presentations after which the reasonable people left quietly (only moderately convinced) and a few not-so-reasonable people approached me in paranoid fashion, wanting to discuss a wide range of topics and approaches that struck me as 'way over the top.'

This is a very odd and momentous time in United States history. There are political, economic, and sociocultural battles across a wide swath of disciplines and activities. American-style capitalism is under siege by global forces that threaten to narrow every nation's ability to grow its economy. American-style capitalism is based on potential and actual growth. Factors that appear to limit the potential for growth, or that actually constrain it, seem to be almost un-American.

Leadership by the United States in the development of technologies has long been at the forefront of both the country's ability to produce economic growth and the high levels of its domestic prosperity. Despite the continuance of social inequities along race, gender, and class lines, the United States and its high technology have long stayed ahead of the curve, leading the world in productivity and standard of living. Clearly, the United States is not the sole world power with the ability to develop innovative technologies. Likewise, it is no longer alone in the world as an economic leader; demographics in China and India, not to mention many other developed and developing countries, long ago put an end to American hegemony over the global economy.

In the spirit of technological achievement, however, the collaborations between the US government, its military, post–World War II intelligentsia, and technology companies led to the development of computational communication-based technologies that changed the equation across almost every segment and sector of economic and everyday life, here and abroad. The rise of computers, computer networks, the Internet, and technologies within the so-called 'information revolution' have introduced numerous socioeconomic complexities.

On the one hand, we could list almost unlimited advances that have increased opportunities for citizens, businesses, and governments to improve every aspect of society. When one laments potential drawbacks of the Internet, robotics, mobile communications, and the like, one risks sounding archaic and faces dismissal from all corners. Who in their right mind would be willing to go without credit cards, mobile/cell phones, the World Wide Web, the Internet, GPS-based technologies, robotics, and digital television, film, and radio, along with an almost endless list of other modern conveniences? The Internet of Things/Everything (IoT) promises to interconnect limitless numbers of devices, to produce uncountable savings, while producing untold levels of productivity and growth. Why would anyone want to question the preferable and inevitable? Doesn't everyone know that there is no reasonable resistance to the pace of technological advances? It is not for nothing that we call critics 'Luddites' and scoff at their unwillingness to go bravely into the future.

There is of course a fly in the ointment—there always is; the devil is always in the details. Simply put, digital leaves tracks. There's no debating the physics of it: The X's and O's in programming code are fairly easily tracked and copied, stored and shared, analyzed and bought/sold. Every single device and every single procedure involved in 'new media' or the 'information revolution' or the 'digital economy' is designed to be machine-readable; after all, that is the point of programming and at the heart of the value of computerization.

This is all a roundabout way of saying that privacy and surveillance together is the single most important subject in relation to American-style democracy.

Very little of the 'American way of life' functions adequately under surveillance. We should not have to be reminded that our fellow citizens most under surveillance are in prisons. Almost everything about being a citizen of the United States, under the sign of democracy, requires privacy. Generations of Americans hardly thought about it; privacy was taken for granted and was only rarely violated. For example, for a few decades in the early years of telephone service, Americans put up with a telephone operator making their connections. These operators had access to the ongoing conversations and could route or reroute calls so that unwanted third parties could listen in. Even though only one person might have listened in, Americans did not tolerate that arrangement for long. Eventually, they considered the practice to be intrusive and demanded that the phone companies develop private-line systems. As hesitant as Americans are today to have or use hardwired phones at home—most would rather use their mobile phones—I would wager that if they returned to hardwired service, they would not consider going back to a party-line system. The notion that others could listen would seem intrusive.

Sometimes, we arrest, prosecute, and jail people who inappropriately view and/or record the private actions of others. We have laws against making recordings of people at the workplace, in public restrooms, and in private changing rooms, and that punish strangers for peeping into our windows at home. We vote in a private booth, we pray in churches without recording our attendance. In effect, it often seems that citizens of the United States have a fairly clear sense of the importance of privacy. In some ways, it seems that Americans remember that their Revolutionary ancestors went to war against the British because those patriarchs from across the sea forced their way into our homes and everyday lives. Most Americans, when confronted with an inability to gather with associates, speak freely, or practice their religion freely, are concerned that their democracy is being compromised.

Americans seem still to know that privacy is important and that surveillance is troublesome. Unfortunately, however, they do not act like they understand or care about either.

The new information and technology economy has encouraged Americans to turn their backs on the value of privacy and to ignore the dangers of surveillance. We are accustomed to the conveniences of a technologized society. We have been unwilling to demand that our elected officials be vigilant. We have been bamboozled by claims that services are free (and apparently without consequences). A government that is deeply involved in surveillance of its own citizens has lied to us. We have allowed fears of enemies and terrorists to cause us to forget the real parameters of a free society.

Commercial films are often an interesting harbinger of our greatest fears. Films sometimes prey on our fears, reflect our subconscious concerns, and appeal to our sense of impending danger. I'm struck by how very often our popular films and television shows represent our fears of totalitarianism. Regardless of whether the invaders are from earthly, yet foreign, soil or are extraterrestrial in derivation, a ragged group of survivors always seeks to regain lost American ground and to reestablish our way of life. And whatever actions these patriots take, they are generally supported by mass audiences who applaud their heroic fight against overwhelming surveillance and the undeniable loss of privacy.

Then, after leaving the theater, folks pull out their mobile phones, tap in a few text messages, make a phone call, go across the street and pay for an ice cream cone with their credit cards, get into their computer chip–laden, partially robotic automobiles, use their GPS navigation systems to find their way home, sit down in front of a computer or digitally enabled television, and tune in to some (more) programming from service providers to which they have subscribed and provided information about themselves. Every one of these steps takes place under surveillance without either the legal expectation or the everyday reality of a shred of privacy.

People argue over whether Edward Snowden is a hero or a traitor. From one point of view, Snowden tried at first to be a whistleblower and then, when blocked and frustrated, gave up his freedom, in support of American democracy, by exposing the bad behavior of our government and our technology companies. From the other perspective, Snowden is a traitor for having turned over an uncountable number of state secrets to an untold number of enemies. I do not know how Snowden will be viewed (or charged) in either the short or long term. I do know, however, that each of us now must face up to the same dilemmas and choices that challenged him.

I fear that the sky *has* fallen and that privacy *is* dead. I fear that I am one of a relatively small number of serious thinkers who believe that our democracy is at risk. I'm an exceedingly minor player in this drama. As an educator in new media technologies, mostly in terms of theory and criticism, I bear some responsibility for helping bring young technologists into the working world without adequately arming them against the forces and choices that they will have to face and make. I bear some guilt over this. Although I have tried to teach about these issues, I have often done so ineffectively and so have contributed to the professionalized cadre of technology workers who do not sufficiently question and challenge their employers about invading citizens' privacy and collecting and selling information into the data marketplace.

Gratefully, I am not alone in my concerns. Increasing numbers of functionaries within the new media environment are raising alarms about privacy and surveillance. For example, *Washington Post* reporter Joel Achenbach, in "Meet the Digital Dissenters: They're Fighting for a Better Internet," recounts a number of critiques:

> Techno-skeptics, or whatever you want to call them—"humanists" may be the best term—sense that human needs are getting lost in the tech frenzy, that the priorities have been turned upside down. ... Of the myriad critiques of the computer culture, one of the most common is that companies are getting rich off our personal data. Our thoughts, friendships and basic urges are processed by computer algorithms and sold to advertisers. ... That information is valuable. A frequent gibe is that on Facebook, we're not the customers, we're the merchandise. Or to put it another way: If the service is free, you're the product. ... Other critics are alarmed by the erosion of privacy. The Edward Snowden revelations incited widespread fear of government surveillance. (Achenbach 2015)

Without wearing a pointed hat made of aluminum—to ward off surveillance and signals—and living in a cave, I write this book in a serious attempt to join with others to raise consciousness and to sound the alarm. Marshall McLuhan often remarked that gaining awareness of the effects of new media is a very difficult endeavor. Generally, by the time folks understand new media well enough to be able to be concerned about them, society is immersed to the degree that citizens can no longer see or care about media effects.

We are, indeed, in over our heads. Whether we can rise to the surface and see clearly is very difficult to know. But the stakes could not be higher. Democracy as it has been known in the United States is at mortal risk due to our loss of privacy and the increase in surveillance. It is time to wake up, to take action, to stop business as usual, and to reshape our culture. I fear it's too late, but I pray that it's not.

INTRODUCTION

We have known, for a very long time, about the foundational privacy principles in the era of computational data management. In relation to the collection, storage, and dissemination of electronically gathered information and records, fundamental objectives and procedures for protecting citizens' privacy have been articulated with authority and regularity at least since 1973.

In that year, the US Department of Health, Education, and Welfare (HEW) issued a report by the Secretary's Advisory Committee on Automated Personal Data Systems titled *Records, Computers, and the Rights of Citizens*. At a time when electronic records were beginning to dominate the governmental scene, the report provided guidelines for federal agencies' handling of electronic records and databases. The report cited a number of basic and definitional aspects of privacy:

> Privacy is the claim of individuals, groups, or institutions to determine for themselves when, how, and to what extent information about them is communicated to others. … [T]his is the core of the "right of individual privacy"—the right of the individual to decide for himself, with only extraordinary exceptions in the interests of society, when and on what terms his acts should be revealed to the general public. (Westin 1967: 7, 373, quoted in *Records, Computers* 1973: sec. 3)

> The right to privacy is the right of the individual to decide for himself how much he will share with others his thoughts, his feelings, and the facts of his personal life. (*Privacy and Behavioral Research* 1967: 8, quoted in *Records, Computers* 1973: sec. 3)

> As a first approximation, privacy seems to be related to secrecy, to limiting the knowledge of others about oneself. This notion must be refined. It is not true, for instance, that the less that is known about us the more privacy we have. Privacy is not simply an absence of information about us in the minds of others; rather it is the control we have over information about ourselves. (Fried 1968, 482, quoted in *Records, Computers* 1973: sec. 3)

The report articulated the Fair Information Practices (FIPs) for electronic privacy that appear, in one form or another, in many subsequent analyses and articulations. These five foundational principles are:

1. There must be no personal-data record-keeping systems whose very existence is secret.
2. There must be a way for an individual to find out what information about him is in a record and how it is used.
3. There must be a way for an individual to prevent information about him obtained for one purpose from being used or made available for other purposes without his consent.
4. There must be a way for an individual to correct or amend a record of identifiable information about him.
5. Any organization creating, maintaining, using, or disseminating records of identifiable personal data must assure the reliability of the data for their intended use and must take reasonable precautions to prevent misuse of the data. (*Records, Computers* 1973: "Summary and Recommendations," "Safeguards for Privacy" [original bullets replaced with numbers])

The report further specified that "[t]hese principles should govern the conduct of all personal-data record-keeping systems. Deviations from them should be permitted only if it is clear that some significant interest of the individual data subject will be served where some paramount societal interest can be clearly demonstrated; no deviation should be permitted except as specifically provided by law" (*Records, Computers* 1973: "Safeguards for Privacy").

Material presented in this book, as well as everyday experience, clearly indicates that little found in these principles pertains to today's commercial data marketplace. Further, the government, in matters of law enforcement and/or security-related surveillance, often does not systematically follow the principles. And yet, the principles have been rearticulated in many important statements about information management since their initial appearance.

For example, the US government enshrined the spirit of the FIPs into law via the Privacy Act of 1974. The Act states that its policy objectives are:

1. To restrict <u>disclosure</u> of personally identifiable records maintained by agencies.
2. To grant individuals increased rights of <u>access</u> to agency records maintained on themselves.
3. To grant individuals the right to seek <u>amendment</u> of agency records maintained on themselves upon a showing that the records are not accurate, relevant, timely, or complete.
4. To establish a code of *"fair information practices"* that requires agencies to comply with statutory norms for collection, maintenance, and dissemination of records. ("Overview of the Privacy Act" 2015)

The importance of the FIPs was further validated by the European Parliament. In a 1995 directive, the EU reminded the world that "data-processing systems are designed to serve man; whereas they must, whatever the nationality or residence of natural persons, respect their fundamental rights and freedoms, notably the right to privacy, and continued to economic and social progress, trade expansion and the well-being of individuals" (*Directive 95/46/EC* 1995: 1–2).

As is often the case, leaders of the EU countries placed the protection of their citizens before the protection of commerce and profit, noting that "the object of the national laws on the processing of personal data is to protect fundamental rights and freedoms, notably the right to privacy" (*Directive 95/46/EC* 1995: 4). The directive stated a number of principles that mirror those found in the HEW statement and in the Privacy Act:

Chapter 1: General Provisions

Article 10
Information in cases of collection of data from the data subject

Member States shall provide that the controller or his representative must provide a data subject from whom data relating to himself are collected with at least the

following information, except where he already has it:
(a) the identity of the controller and of his representative, if any;
(b) the purposes of the processing for which the data are intended;
(c) any further information such as
—the recipients or categories of recipients of the data,
—whether replies to the questions are obligatory or voluntary, as well as the possible consequences of failure to reply,
—the existence of the right of access to and the right to rectify the data concerning him in so far as such further information is necessary, having regard to the specific circumstances in which the data are collected, to guarantee fair processing in respect of the data subject.

Article 12
Right of access

Member States shall guarantee every data subject the right to obtain from the controller:
(a) without constraint at reasonable intervals and without excessive delay or expense:
—confirmation as to whether or not data relating to him are being processed and information at least as to the purposes of the processing, the categories of data concerned, and the recipients or categories of recipients to whom the data are disclosed,
—communication to him in an intelligible form of the data undergoing processing and of any available information as to their source,
—knowledge of the logic involved in any automatic processing of data concerning him ...
(b) as appropriate the rectification, erasure or blocking of data the processing of which does not comply with the provisions of this Directive, in particular because of the incomplete or inaccurate nature of the data;
(c) notification to third parties to whom the data have been disclosed of any rectification, erasure or blocking carried out in compliance with (b), unless this proves impossible or involves a disproportionate effort.

Article 14
The data subject's right to object

Member States shall grant the data subject the right:
(a) ... to object at any time on compelling legitimate grounds relating to his particular situation to the processing of data relating to him, save where otherwise provided by national legislation. Where there is a justified objection, the processing instigated by the controller may no longer involve those data;
(b) to object, on request and free of charge, to the processing of personal data relating to him which the controller anticipates being processed for the purposes of direct marketing, or to be informed before personal data are disclosed for the first time to third parties or used on their behalf for the purposes of direct marketing, and

to be expressly offered the right to object free of charge to such disclosures or uses. (*Directive 95/46/EC* 2, 10)

In short, fundamental principles that were, at one time, goals and ideals in the United States are now foundational throughout the EU. In addition, the EU version specifies increased levels of protection for private citizens.

One need only look to 2009 to find an example of the powerful righteousness of the FIPs. In February, staff of the Federal Trade Commission (FTC) issued a report titled "Self-Regulatory Principles for Online Behavioral Advertising" that rearticulated principles based on the FIPs. A mere six months later, five important US trade associations (American Association of Advertising Agencies, Association of National Advertisers, Council of Better Business Bureaus, Direct Marketing Association, Interactive Advertising Bureau) based a series of recommendations on the FTC report, to the point of replicating its title—*Self-Regulatory Principles for Online Behavioral Advertising*—and adhering to the same FIPs. These five organizations, all with extensive commercial interests in collecting user data, joined the FTC in calling for increased education of consumers about online behavioral advertising along the lines articulated in the FIPs.

One can criticize the reports for targeting consumers (with so-called education) instead of disciplining industry operatives and practices. While the title *Self-Regulatory Principles for Online Behavioral Advertising* implies changes to industry practices, the reports propose that industry participants accomplish those changes by educating the public about how online behavioral advertising implements and uses data collection. Nevertheless, the principles articulated as the goals of these educational efforts echo those represented in HEW's FIPs, the Privacy Act, and EU directive lists. In the industry report, these principles are outlined as follows:

> The Transparency Principle requires the deployment of multiple mechanisms for clearly disclosing and informing consumers about data collection and use practices associated with online behavioral advertising. ...

> The Consumer Control Principle provides for mechanisms that will enable users of Web sites at which data is collected for online behavioral advertising purposes the ability to choose whether data is collected and used or transferred to a non-affiliate for such purposes. ...

> The Data Security Principle requires entities to provide reasonable security for, and limited retention of, data collected and used for online behavioral and advertising purposes. ...

> The Material Changes Principle directs entities to obtain consent before applying any change to their online behavioral advertising collection and use policy that is less restrictive to data collected prior to such material change. ...
>
> The Sensitive Data Principle recognizes that certain data collected and used for online behavioral advertising purposes merits different treatment. The Principles apply heightened protection for children's data by applying the protective measures set forth in the Children's Online Privacy Protection Act. Similarly, this Principle requires consent for the collection of financial account numbers, Social Security numbers, pharmaceutical prescriptions, or medical records about a specific individual for online behavioral advertising purposes.

In the face of the ubiquitous and ever-burgeoning data marketplace and government surveillance programs, numerous contemporary political leaders answer privacy challenges by returning to the FIPs. In the 2012 report *Consumer Data Privacy in a Networked World: A Framework for Protecting Privacy and Promoting Innovation in the Global Digital Economy*, the Obama administration presented a proposal for a Consumer Privacy Bill of Rights featuring the following FIP-based aspects:

> Individual Control: Consumers have a right to exercise control over what personal data companies collect from them and how they use it.
> Transparency: Consumers have a right to easily understandable and accessible information about privacy and security practices.
> Respect for Context: Consumers have a right to expect that companies will collect, use, and disclose personal data in ways that are consistent with the context in which consumers provide the data.
> Security: Consumers have the right to secure the responsible handling of personal data.
> Access and Accuracy: Consumers have a right to access and correct personal data in usable formats in a manner that is appropriate to the sensitivity of the data and the risk of adverse consequences to consumers if the data is inaccurate.
> Focused Collection: Consumers have a right to reasonable limits on the personal data that companies collect and retain.
> Accountability: Consumers have a right to have personal data handled by companies with appropriate measures in place to assure they adhere to the Consumer Privacy Bill of Rights. (*Consumer Data Privacy* 2012: 1)

Once again, there is very little equivocation as to how the FIPs are constituted, as this presentation mirrors previous articulations going back to the HEW document of 1973. However, significant differences exist in historical contexts. In 1973, the government articulated principles for the data management practices it planned to develop for a time (in the future) when electronic federal databases and records would become functional realities. In 2012, the government rearticulated the principles, fully knowing that few (if anyone) in either government or the private sector *ever* follows the FIPs within environments where data collection, exchange, and commerce have joined with surveillance to become multi-billion-dollar industries at the very heart of the Internet age.

In some ways, this book is driven by a very simple question: Given how much we have known, for a very long time, about new media's threats to our privacy, why on earth have we capitulated and given up a fundamental right so quickly, thoroughly, and easily? I have often speculated with students in classes—somewhat hyperbolically, to be sure—that had US citizens in my parents' generation (or earlier) been faced with intrusions on their property rights (real estate) at levels approaching our contemporary loss of privacy, Americans would have staged a second revolution or civil war.

Evidence presented in this book indicates that within a relatively short period—less than 20 years—a previously taken-for-granted and highly valued right (privacy) has eroded, or been stripped, to an alarming degree and with numerous detrimental implications. And yet, most Americans go blithely along, as though nothing untoward has happened, in spite of frequent press reports about the troubles. Most organizations, be they governmental or private, for-profit or not-for-profit, participate (mostly fully) in the data marketplace and, to a very large extent, ignore the FIPs or skirt them via privacy policies, terms of service (ToS), end-user license agreements (EULAs), contracts, or notices—almost always in ways that citizen-consumers don't read, don't understand, and/or aren't able or willing to refuse. It is as if slavery, prohibition, and the Flat Earth Society have returned to prominence, perhaps even dominance, yet the populus act as though everyday life is unaffected.

I've long wondered, how could this happen? How did we get to this point, especially given our initial high expectations for progressive outcomes from the digital revolution? The reasons are many, complex, and hotly contested, across a wide variety of vested interests. A brief historical review orients further discussion.

How the Heck Did We Get Here?

Lest I seem overly naïve, I should begin by noting some obvious facts. First, there have always been spies and surveillance, whether employed by the military, FBI,

CIA, or NSA. Second, law enforcement has long relied on surveillance techniques. At times, the methods used have crossed the lines of legality and propriety. Third, since the turn of the 20th century, laws and courts have recognized civil privacy claims between persons because citizens sometimes invade each other's privacy. The Constitution speaks to many of the protections surrounding privacy and surveillance, although some argue that our laws no longer protect us adequately. Fourth, corporations and business interests have long made great efforts to reach potential customers/consumers with advertising and marketing. The sophistication of their techniques has developed and improved over time, and advertising and marketing have long been important aspects of capitalism.

As noted, at the point in history when computerized databases began to take a prominent role in the collection and storage of information, agencies of the US government looked to develop practices that would constrain surveillance and control data management. Likewise, legislative efforts were taken to control the ways in which government agencies and bodies collect and exchange information about citizens.

Although one cannot point to a hard and fast line of demarcation, it is likely that many citizens prior to the year 2000 might have said that the checks and balances over privacy and surveillance issues were adequate in the United States. However, privacy and surveillance experts were beginning to doubt the adequacy of those checks and balances and controls. Participants in, and theorists of, new media activities began issuing warnings concerning the way the data might be managed. So what went wrong? What changed? Any discrete listing of features is oversimplified. However, certain events and activities served as watershed habits and moments.

Due to changes in the economy and consumer habits, the United States has seen broad increases in the use of consumer credit, largely in the form of credit cards. Moving away from cash and using credit cards produces an electronic trail.

The development of mobile telephony and of the functionalities found in cell/mobile phones and smartphones increases exponentially the ability to track the location of individuals and to collect data from their electronic communications.

Obviously, the Internet facilitates—or need we say, thrives on—technologies that find citizens connected (most waking hours) to networks that produce tracking records. While the early Internet, sometimes referred to as Internet 1.0, did not produce extensive and accurate tracking, later, more interactive, versions—2.0 and beyond—do so.

Although data management and protection practices and privacy protection laws have been passed as electronic technologies developed, these constraints generally apply only to governmental entities, and more particularly, mostly at the federal level. Precious little privacy legislation constrains private enterprise. Indeed, once private enterprise decided to leverage information about people into targeted marketing, the entire landscape of privacy and surveillance changed radically because there are few legal constraints on the private data marketplace.

Related to this feature, aspects of social political systems in the United States that are supposed to protect citizens have instead shifted to a priority for corporations and industries. New media industries are especially fortunate in their reception of this benevolence. Legislators and courts have treated Internet-related activities with kid gloves. At the beginning, legislators and courts did not want to damage the new industries as they developed, so, for example, taxation was limited as it related to networked Internet commerce. Courts have long found that new industries sometimes need extra protection; thus, interpretations of how contract law works on the Internet, for instance, are much less restrictive than in terrestrial industries.

However, neither the courts nor legislatures have changed their approaches to these industries very much, even long after new media stopped being young, frail, and vulnerable. Even as the Internet giants have become behemoths, legislation and court decisions have failed to provide adequate controls. In fact, the Internet remains relatively unregulated.

Technological achievements are not limited to human communication aspects. Many technological developments have taken place in machine-to-machine communication. For example, cars, appliances, utilities, and other devices contain computerized technologies that track, record, and transmit reports about operation and usage.

The attacks that occurred on 9/11 changed everything. Subsequent legislation, particularly the Patriot Act and related national security practices and legislation, overrode most extant protections against government surveillance practices. As a result, government surveillance agencies—the National Security Agency (NSA), in particular—increased their size and power to an extent that was previously unimaginable. In fact, practices that legislatures and citizens had previously, and strongly, opposed became de rigueur in the post-9/11 climate of fear.

Citizen participation in social media increased exponentially, with Americans using social media platforms such as Facebook, Twitter, Instagram,

Foursquare, and YouTube. In fact, ordinary citizens probably provide more identity-specific information to and through social media than is collected by surveillance agencies—and they do so freely and, in many cases, enthusiastically. It's sometimes difficult to argue that government or new media industries should do more to protect privacy when so many subjects of surveillance provide information about themselves freely, as though they do not care to protect their own privacy.

One of the important and recurring themes in considerations of how privacy and surveillance should work in the United States has to do with the relationship between public and private streams of data. To a certain extent, citizens want the government to have and protect information that enables governmental functions. For example, citizens want the social security and driver licensing systems to work, they want to be able to vote, and so on. Citizens also realize that the government will have to engage in some degree of surveillance and data collection if it is to protect citizens from terrorist attacks, attacks by other enemies, or lawlessness. Although citizens are often concerned that the government overreaches its authority in these areas, the expectation exists that government must engage in these activities at some level, along with the hope that safeguards against government misbehavior are in place. In part, this explains why citizens seem to accept some level of intrusion on their private lives. It also accounts for many citizens' sense that "I'm not worried about government surveillance because I've done nothing wrong." This attitude represents the idea that the government's activities are surely limited to surveillance of criminals or potential terrorists: our enemies. Most citizens don't think of themselves as members of those categories and, in turn, don't think they are targets of surveillance. And even if they are, the surveillance is not picking up anything illegal or worrisome.

We will return to this later; clearly, the Snowden revelations compromise some of these naïve ideas.

An additional and important recurring theme about privacy and surveillance has to do with advertising and marketing. Citizens of the United States tend to believe that marketing and advertising are a social good and should generally be allowed to take place without government intervention. Further, and somewhat informally, Americans don't think much of marketing and advertising. That is to say, they don't believe they are influenced much by it: "It's just advertising—I ignore most targeted marketing anyway." While most Americans report that they don't want to be tracked and they don't like targeted marketing, they're willing to put up with it as part of the way the economy works.

Privacy problems get more serious when the two data streams, commercial and governmental, overlap. One of the principle reasons that Americans give the government and private industries free passes to collect data is that they assume that the two sides are not in cahoots. However, this expectation no longer comports with reality. Government and private enterprise now share data regularly. In fact, virtually everyone who collects data participates in the data marketplace, such that governmental and private data streams are no longer separate. The loss of privacy is the result, as surveillance increases and personal data is distributed throughout the data marketplace.

The justifications, reasons, and excuses include many perceived or potential tangible benefits in the face of an enormous range of hidden costs. Organizing these costs and benefits into categories provides a preview of important topics in this book.

Technological Complexities in Data Collection: Out of Sight, Out of Reach

Most data collection procedures take place outside of users' conscious awareness and deliberate control. According to a 2014 Pew Research Center study and report, "2,558 experts and technology builders ... foresee an ambient information environment where ... [i]nformation sharing over the Internet will be so effortlessly interwoven into daily life that it will become invisible, flowing like electricity, often through machine intermediaries" (Anderson and Rainie 2014).

Contemporary users utilize many devices and a wide variety of accounts. Managing privacy for each device and account requires multiple procedures. Even sophisticated users find the proper monitoring and management of privacy settings to be difficult and time consuming. The average, less sophisticated user is either overwhelmed or simply unable to effectively manipulate settings properly. Further, in many instances, setting privacy more securely results in limitations to, or loss of, wanted services.

The people who implement data collection procedures tend to be extremely sophisticated, especially in comparison to the average user. Many information management employees are tasked with implementing practices that take advantage of users' data; doing so is a specific function of their job and, as such, is very difficult for them to resist. If and when these employees do their jobs well, the everyday user finds it difficult to push back against such implementations.

Data collection has become part and parcel of how online commerce, communication, and activities in general work. Most participants in the industries involved have made ongoing information management procedures de rigueur. Overturning them seems impossible and/or inadvisable from a corporate perspective.

The amount and nature of tracking has changed radically as, in addition to collecting data about online activities, "online marketers are increasingly seeking to track users offline ... by collecting data about people's offline habits—such as recent purchases, where you live, how many kids you have, and what kind of car you drive" (Angwin, "Why Online Tracking" 2014). Julia Angwin quotes Acxiom's chief executive of broker firms, Scott Howe, who notes that "The marriage of online and offline is the ad targeting of the last 10 years on steroids." Angwin notes that while "companies that match users' online and offline identities generally emphasize that the data is still anonymous because users' actual names aren't included in the cookie ..., critics worry about the implications of allowing data brokers to profile every person who is connected to the Internet" (Angwin, "Why Online Tracking" 2014).

An investigative report published in the *Wall Street Journal* in 2010 noted that the business of spying on Web users is one of the fastest growing businesses on the Internet. Examining the surveillance technology that companies are deploying on Internet users,

> [t]he study found that the nation's 50 top websites on average installed 64 pieces of tracking technology onto the computers of visitors, usually with no warning. A dozen sites each installed more than a hundred. ... Tracking technology is getting smarter and more intrusive. ... [T]he Journal found new tools that scan in real time what people are doing on a Web page, then instantly assess location, income, shopping interests and even medical conditions. Some tools surreptitiously re-spawn themselves even after users try to delete them. These profiles of individuals, constantly refreshed, are bought and sold on stock-market-like exchanges that have sprung up in the past 18 months. ... In between the Internet user and the advertiser, the Journal identified more than 100 middlemen—tracking companies, data brokers and advertising networks. ... The Journal examined the 50 most popular U.S. websites, which account for about 40% of the Web pages viewed by Americans. ... As a group, the top 50 sites placed 3,180 tracking files in total on the Journal's test computer ... over two-thirds—2,224—were installed by 131 companies, many of which are in the business of tracking Web users to create rich databases of consumer profiles that can be sold. (Angwin 2010)

Thomas Allmer et al. note that most social media users are aware that site operators collect and store personal information for targeted advertising. However, the authors remind readers that

> there is a great lack of knowledge when it comes to details about the actual process of data collection, storage and sharing. Respondents of our study were quite uncertain or even misinformed about what exactly Facebook is allowed to do with their personal data and which personal data, browsing data and usage data is actually used for the purpose of targeted advertising. This may partly be explained by the fact that privacy policies and terms of use are often lengthy, complicated and confusing. (Allmer et al., 2012: 65)

Applications, networks, services, platforms, and companies that collect user data as an everyday practice are so ubiquitous and embedded that it is virtually impossible for users to avoid them. 'Dropping off the grid' is next to impossible, short of moving to a cave in a remote location.

The degree to which surveillance technologies—information/data collection, sharing, marketing, management—operate in the deep background plays a strong role in the seeming inability of everyday citizens to care about and/or control access to their private information. Digital services providers have made enacting privacy constraints very difficult, both to understand and to execute. Controls for the potential actions that users might want to take are embedded so deeply in the ways the technologies operate that the average person simply cannot or will not invest the time and energy to find them, manipulate them, and control them. Often, doing so ruins the mediated experience or blocks one from being able to participate at all.

Two recent books are particularly instructive regarding the technological complexities involved when users attempt to take charge of their personal information by limiting the abilities of data collection entities.

In her book *Dragnet Nation*, journalist Julia Angwin (2014) painstakingly describes the procedures she used in order to evade surveillance. Over the course of a year, Angwin created a false identity so she could pollute the accuracy of the data market analyses resulting from her purchases and website use. She used sophisticated password generators to change her logon protocols, learned to use encryption software, carried a wallet that blocked radio-frequency identification signals, and opted out of a large number of data collection services (and failed to opt out of many others), among a wide range of other actions. After a year, and toward the end of the book, Angwin reported feeling "surprisingly hopeful":

On one level, my efforts to evade the dragnets were not very successful. I hadn't found a way to use my cell phone—or my burner phone—in a way that protected my location and calling patterns. ... I hadn't extricated myself entirely from the clutches of Google and Facebook. My name and address were still on file at more than 100 data brokers that didn't provide me with a way to opt out. And I wasn't going to be able to avoid facial recognition cameras. But, on another level, I had exceeded my expectations. I had avoided the vast majority of online ad tracking. My passwords ... were pretty good. My fake identity ... [a]llowed me to disassociate my true identity from sensitive purchases and some phone calls and in-person meetings. And I had managed to convince some of my friends and sources to exchange encrypted texts, instant messages, and emails. (*Dragnet* 2014: 210)

Angwin's expertise, motivation, and efforts far exceeded levels available to normal citizens. Her position as a journalist on the digital beat provided her with a wide range of expert consultants. Her efforts were aided and abetted by her publication goals: articles for newspapers and magazines and material for the book. Yet even though much of her focus across an entire year was on the evasion of surveillance and data collection, she was unable to control many of those intrusions and certainly unable to 'drop off the grid.'

Perhaps an even more compelling instance is provided by Michael Bazzell (2013), in his book *Hiding from the Internet*. As a security expert, Bazzell is even less a common citizen than is digital journalist Angwin, having served in law enforcement and then worked for the FBI's cyber crimes task force. The 10 chapters in his book include strategies such as completing insanely thorough self-background checks with hundreds of data collection entities. Angwin (*Dragnet* 2014) found this phase of the process infuriatingly frustrating due to the high number of repetitive steps that one must go through in order to ascertain what material is held by literally hundreds of data collection entities—and how to opt out.

Bazzell then describes how to set up anonymous email accounts; manipulate the image on one's driver's license, to prevent accurate facial recognition by data collection firms; manage one's browser and browser plug-in settings; remove oneself from social media; delete inaccurate information that has already been gathered; opt out of data collection from targeted marketing networks; check one's credit reports—in short, how to remove as much information about oneself as possible and block further collection of personal data.

Bazzell's treatment is thorough, and the steps he provides are all within the reach of most computer-savvy users. However, the time and effort required to complete even a few of the steps he recommends, and thereby "hide from the Internet," are far beyond what most normal citizens would be willing and/or able to engage in.

In short, one takes away from these two books the idea that protecting personal information has become far too difficult for normal users and requires extraordinary expertise and measures. The situation narrows the action space available to citizens and discourages them from taking any action. It all seems so overwhelming and so complicated; the steps required take so much time and effort that it is often just *easier* to allow surveillance and ubiquitous data collection. Allowing surveillance and data collection seem to be the only viable options for reasonable people who want to use digital resources and feel that they derive some benefits from doing so.

Product and service providers take refuge behind notifications of ToS and privacy policies. However, these provide little (if any) real privacy protection for consumers. By providing links to documents titled "privacy policies" and "terms of service," corporate entities give the appearance of transparency. However, the terms and policies are difficult for users to comprehend and are often unreasonable or overreaching. Further, the links and materials often connect to additional documentation and other providers, and the controls provided are confusing and relatively ineffective in the hands of the average user. More than anything, most of the activities controlled by terms and policies (i.e., the actual data collection and distribution) take place out of view to the extent that users seldom think about these activities; when they do consider them, the degrees of difficulty and the potential for (or reality of) service degradation block users' efforts to control their personal data.

These reasons, accounts, explanations, and excuses are so common as to be ubiquitous within the cultural conversation in the United States. Over the past five years literally thousands of articles have been published by print and online news outlets and read by everyday citizens. Additionally, a plethora of magazines, blogs, and trade-focused punditry directed toward digitally savvy users focuses on issues of privacy and security with everyday regularity in their sections, columns, and articles. How on earth can this much information about data management (privacy, security, and the data marketplace) go virtually ignored by the vast majority of digital users?

In his book *The Glass Cage*, Nicholas Carr (2014) instructs and reminds us about the rapidity with which new technologies become commonplace. This idea was articulated by Marshall McLuhan, often through a metaphor noting that dominant media become like water to fish: By the time one is fully immersed in a dominant new medium, one is unlikely to notice it any more than fish take note of their water or than we take notice of the air that we breathe (McLuhan and Fiore 1967). Only when a polluted environment begins killing off the inhabitants does one notice the filth, and by that time it's too late to turn back. Carr writes

> If we don't understand the commercial, political, intellectual, and ethical motivations of the people writing our software, or the limitations inherent in automated data processing, we open ourselves to manipulation. ... The more we habituate ourselves to the technology, the greater the risk grows. ... Even when we are conscious of their presence in our lives, computer systems are opaque to us. Software codes are hidden from our eyes, legally protected as trade secrets in many cases. Even if we could see them, few of us would be able to make sense of them. They're written in languages we don't understand. The data fed into algorithms is also concealed from us, often stored in distant, tightly guarded data centers. We have little knowledge of how the data is collected, what it's used for, or who has access to it. Now the software and data are stored in the cloud, rather than on personal hard drives, we can't even be sure when the workings of systems have changed. ...The modern world has always been complicated. Fragmented into specialized domains of skill and knowledge, coiled with economic and other systems, it rebuffs any attempt to comprehend it in its entirety. But now, to a degree far beyond anything with experience before, the complexity itself is hidden from us. It's veiled behind the artfully contrived simplicity of the screen, the user friendly, frictionless interface. We're surrounded by what the political scientist Langdon Winner has termed "concealed electronic complexity." ... When an inscrutable technology becomes an invisible technology, we would be wise to be concerned. At that point, the technology's assumptions and intentions have infiltrated our own desires and actions. We no longer know whether the software is aiding us or controlling us. (Carr 2014: 213–214)

Appealing to Consumers' Lowest Common Denominators: It's All Good?

In general, especially with regard to technology, American citizens tend to be somewhat lazy, preferring technologies to be easy to use, entertaining, and frictionless (Reeves and Nass 1996).

Many consumers derive perceived or actual benefits from digital technologies. Often, though, consumers fail to perceive, understand, or appreciate the real costs when exchanging the information (personal data) required for those perceived or actual benefits. Informed cost-benefit analysis is not possible when actual costs are hidden and/or unavailable. However, the benefits, perceived or actual, are tangible and so appear to outweigh any hidden or unknown costs. Information does not really 'want to be free,' and 'free' services don't usually come without costs.

Posting to Facebook and Instagram, posting to and following on Twitter, reading online newspapers, reading and posting to online blogs, posting product reviews, searching for and acquiring products online, sending text messages using computers and mobile phones, posting comments about postings by oneself and others, using credit cards, driving cars with 'smart' technology—these activities have become ubiquitous, all-encompassing parts of the total life experience. Immersion in this much data exchange across such a wide variety of devices and activities makes controlling personal data very challenging. The imminent onset of the Internet of Things/Everything further exacerbates the situation, as an even wider range of devices, networks, and activities will produce exponentially more data about personal matters.

Use of these technologies is ubiquitous. Since 'everyone else' seems to be using them, one risks being left out by not using them. The use of digital technologies relates positively to perceptions of affluence, knowledge, sophistication, and other features of upward social mobility. It also shows off one's toys, abilities, belongingness, and erudition. Often, users improve and/or increase their activities when supported by technological mediation.

Another, even less forgiving, reason exists for why we've allowed the situation to progress as far as we have: We've stopped treating one another as human people and moved a long way toward treating one another as objects and as abstracted forms of data. Vincent Miller's analysis proposes that "virtual matter is conceived of as 'information about' beings as opposed to 'the matter of being' in contemporary environments. This allows aspects of the contemporary self (i.e. data about ourselves) to be treated as commodities, not as meaningful components of the self but as a series of potentially useful or valuable objects (data to be used and sold)" (Miller 2016: 8).

Supporting Economic Interests of Business and Industry: Doubling Down on Data

The economic impacts of the big data economy are enormous in both potential and actual dollar amounts. Billions of dollars change hands in transactions that depend, and in some cases are based on, the data marketplace, involving industries both within and outside of new media.

The digital, new-media/online economic environment transitioned from a less data intensive status to a deeply data intensive status as a way to increase profits. The transition has been highly successful, generally increasing profits, especially for large, corporate, digital interests. Industries involved directly in online commerce, innovation, and development (e.g., Google, Facebook, Amazon, and the big firms that make up the data analysis segment) are unlikely to have much interest in reducing the use of big data and curtailing the resultant profit.

Hundreds of companies participate in the data marketplace functions of collecting, analyzing, packaging, and selling personal data. In contrast to front-facing social media entities that are well known to a public filled with enthusiastic users (Facebook, Twitter, Instagram, Foursquare, and the like), the major players in the data marketplace are less widely known, although they are mentioned often in books, articles, and press reports. These players include Acxiom, Rapleaf, eXelate, BlueKai, and others.

Acxiom "claims it can help marketers by selling demographic data such as age, gender, marital status, race, ethnicity, address, and income … covering over 126 million households and approximately 190,000,000 individuals" (Turow 2011: 97). Acxiom—an international data-gathering giant with annual sales of around $1.1 billion—has over 23,000 servers in Arkansas that are always collecting, collating, and analyzing consumer data (Angwin, *Dragnet* 2014: 87; Bazzell 2013: 71).

Turow highlights other major data marketplace players:

> Rapleaf is a firm that claims on its website to help marketers "customize your customers' experience." To do that, it gleans data from individual users of blogs, Internet forums, and social networks. [It uses] ad exchanges to sell the ability to reach those people. Rapleaf says it has "data on 900+ million records, 400+ million consumers, [and] 52+ billion friend connections." (Turow 2011: 4)

> eXelate says it gathers online data from two hundred million unique individuals per month through deals with hundreds of websites. … eXelate determines a consumer's age, sex, ethnicity, marital status, and profession by scouring website registration. … It gathers information anonymously using tracking cookies placed on the hard drive of the consumer's computer. … Through what eXelate calls its Targeting eXchange, advertisers bid on the right to target consumers with those cookies (as well as those of other firms), and eXelate shares this revenue with the sites that provided access to the data.

> BlueKai, an eXelate competitor, also strikes deals with thousands of websites to use cookies to collect anonymous data about activities. (Turow 2011: 79–80)

Angwin reports that "Epsilom, one of the largest direct marketers, [boasts] more than $3 billion in annual sales" (*Dragnet* 2014: 86). Additionally, Bazzell notes that Intelius has one of the largest personal information databases on the Internet (2013: 75) and LexisNexis, "one of the largest data aggregation companies in existence ... is owned by Reed Elsevier, which is also the parent company of Choicepoint, Accurint, and KnowX" (Bazzell 2013: 73).

A July 17, 2015, visit to the CNN website—using Firefox enabled with the Disconnect plug-in—garnered data requests from 12 different services as well as ongoing requests from over 110 Facebook-generated linkages (I don't have a Facebook account). Since I've blocked these services, we'll call these 'data requests' rather than 'data collection.' Requesters included TRUSTe, Outbrain, Nielsen, ClickTale, Chartbeat, Optimizely, ScoreCardResearch.com, Fonts.net, Visual Revenue, Livefrye, Akamai, Facebook, Twitter, and Google. A quick look at the Privacy Badger plug-in reveals that, in addition to data/information requests, trackers from nine outfits were attempting to load cookies onto my computer: rtax.criteo.com, cdn.optimizely.com, apostrelease.com, aspen.turner.com, i.cdn.turner.com, i2.cdn.turner.com, z.cdn.turner.com, www.ugd.turner.com, and cdn.xrxd.net.

Online, Web-based data collection is only the tip of the iceberg but plays an important role in the provision of data points in the marketplace. It is also one of the most insidious modes. Users browsing websites for news, entertainment, and social connectivity are engaged in activities they would rate as trivial in comparison to, for example, shopping for a home or mortgage, shopping for a car or car loan, receiving medical treatment or purchasing prescriptions, etc. This sense of triviality in much of their Web use leads everyday users to take no significant action to block data collection.

Further, using technological measures to block data collection erodes the Web-browsing experience. Browsing for news, information, entertainment, and social connections is supposed to be quick, easy, and fun; putting technology-blocking software in place tends to make websites 'clunky' and/or the viewing experience incomplete.

Government and law enforcement operatives are able to utilize the private data marketplace in ways that advantage surveillance and law enforcement functions. Government and law enforcement now have a big stake in maintaining the private data marketplace, although they may mask their interest by actions that appear to lightly constrain information management activities and/or forewarn citizens of risks.

US laws often give corporate entities the advantage over private citizens. Many of these laws and practices relate directly to information management and the data marketplace. Structural foundations of law and politics in the United States tend to protect corporations and profits more than private citizens. While this might be expected and preferred by corporate capitalism, it tends to work somewhat strongly against a variety of individual rights and interests.

Unwillingness of Government to Protect Its Citizenry from Itself: Corporations Are Our Most Valued People

Precious little law applies to the private data marketplace in the United States. Generally speaking, the data marketplace developed, and now operates, behind the back of regulation. Although a number of laws, in theory, protect citizens from governmental and law enforcement use of big data, recent world events (the war on terror, global surveillance) and domestic politics have combined to weaken the effectiveness of existing laws to protect the citizenry. Further, efforts to constrain governmental use of big data run the risk of impinging on the private sector's uses and therefore are politically dis-preferred.

Just as legislators in the United States are unwilling to undermine corporate profits and to untangle citizens' information from unknowing and/or unwilling participation in the data marketplace, our representatives have long been unwilling to become involved with intellectual property legislation. Some of the potential constraints for data management could be enacted via changes to intellectual property law, but in the current political environment, such changes are unlikely to happen without interested politicians.

The courts appear to have lost all good sense regarding how contract law applies to digital technologies. Court interpretations of wrap contracts—be they ToS, EULAs, shrink wraps, or click-throughs—appear to have left the realm of the reasonable and entered the kingdom of the unimaginable. As a result, contract law principles that previously protected citizens no longer do so; in fact, they now prey on consumers.

In a way that is similar to the situation of intellectual property law, consumers have little or no advocacy power against corporate or governmental data management practices. Because their elected representatives and much of the legal system work harder to support corporate interests than individual interests, Americans are, in effect, seldom 'at the table' in policymaking discussions regarding their information and data.

Concerns over Terrorism and Security Risks: A Darned (In)Convenient War on Terror

It can certainly be said that the 9/11 attacks changed everything. Most directly and obviously, the US government and its security apparatus embarked on a series of policies and practices that increased surveillance and data collection exponentially. As noted, participation in these important security functions tends to make the US government and law enforcement less interested in curtailing the private data marketplace. As the data streams (public and private) cross, the rights of citizens are reduced.

Digital and computational technologies give law enforcement entities effective tools for pursuing criminals and curtailing illegal activities. One would not expect law enforcement entities to purposefully reduce their use of effective technologies. However, as is the case with any law enforcement apparatus, monitoring from outside entities is required as a counterbalance for the potential or actual misuse of technological tools. Generally speaking, the abilities of law enforcement and government exceed those of private citizens. As a result, technologies that citizens use and treat as social and benign are often used, by law enforcement agencies, against everyday people. Governmental (especially within the intelligence community) and law enforcement entities share in the massive amount of commercially based data collection. Neither the intelligence nor law enforcement communities have much (if any) interest in limiting commercial data and the data marketplace. Politicians playing on, and to, fears of terrorism promote the maintenance and increase of data collection. Everyday users of social media and other digital technologies, who behave as though they don't care about privacy or surveillance, complete the circle. It's very easy to blame it all on the war against terror and the need for heightened security.

Family Observations: A Personal Anecdote

Single anecdotes are never universal indicators. Yet, my own family experiences inform my perspective on the ways that young people treat privacy and surveillance. Our four children, one boy and three girls, now range in age from 20 to 28 years of age. They are millennials and digital natives, having grown up during the development of new media and the Internet.

Although my wife and I put some restrictions on time spent watching television, our children generally had access to communication via electronic devices and the Internet as they grew up. They had less access to sophisticated mobile phones than did most of their peers; we did not purchase smartphones for any of the children, instead providing flip phones without text messaging until the children were out of high school. At that point, three of the four purchased their own cell phones, opting for iPhones (we are an Apple/Mac family).

None of the four entered a media-production industry. Three of the four are moderately sophisticated digital users. All four utilize a fairly wide range of social media. One is an avid player of video games. Generally, their use of digital devices is very similar to that of a large proportion of their generation; 95% of teens use the Internet, 78% have cell phones, and 81% use social networking sites (Rainie 2014).

As is the case with most everyday citizens of the United States, our four adult children do not take special steps to protect their privacy in the digital realm (Rainie and Madden 2015). Despite having a father who has spent longer than a decade railing about privacy issues, the people in our family (my wife included) have no more sophisticated an approach to privacy protection than do most folks. They regularly post to Facebook, post to and follow others on Twitter and Instagram, use location navigation with their cellular devices, and use relatively unprotected email clients.

Polling data indicates that Americans teenagers say that they care about privacy (Madden et al. 2013). However, as indicated by their behavior that we have witnessed as they've grown up, our children act as though they care little about privacy in the digital realm.

In comparison to their online behavior and expectations, our children are excessively demanding with regard to terrestrial personal privacy. I have to assume this indicates that they do care about privacy. In fact, they care about it very much, in the same way that Americans have made privacy a priority for a couple hundred years and that some claim to care about it now.

Our family has always lived in modest, middle-class housing. Throughout most of their childhood and adolescence, our children shared a bedroom with at least one of their siblings; two sometimes had rooms of their own. Eventually, we were able to provide a separate bedroom for each child. Regardless of where we lived or how our children were assigned bedrooms, they screamed, cried, and stomped their way through demonstrations demanding privacy. They wanted locks on doors (that we refused to provide); in addition, they wanted privacy in the bathroom, privacy in the shower area, privacy when they did homework, private places for computer use, and privacy for phone conversations. Their behavior made it clear that even young people place a high value on personal privacy.

Mediating these factors has become excruciatingly difficult. It's easy to get discouraged in the face of the data collection and marketplace onslaught. However, approaches exist that can moderate these forces. As we see in the reactions of citizens in countries other than the United States (Levine 2015), and as we learn from simple anecdotes like these about my family situation, a strong possibility remains that privacy, and wanting it, is alive and well in the United States (and possibly more so overseas). Privacy just isn't easy to attain. A wide range of participants have to cooperate if we are to regain control over our personal data.

Chapter Previews

This volume examines challenges we face in a wide range of contexts including the personal, business, governmental, and societal. It investigates and interrogates our systems of data management, including the ways that data are collected, exchanged, analyzed, and repurposed. The primary focus is the commercial data marketplace; however, the overlapping interests and activities of government surveillance and data collection saturate the overall topic and thus cannot be ignored. The work begins, in chapter 1, by substantiating the severity of the problems at hand and, in doing so, provides evidence to bolster the speculative claims made in this introduction.

Chapter 2 presents conceptual frameworks for interrogating the data management environment, drawing primarily from perspectives offered by Robert McChesney (2013, 2015), Helen Nissenbaum (2010, 2015), Lawrence Lessig (2001, 2006), and José van Dijck (2014). From a broader view, questions posed by Neil Postman (1992, 1999) toward technologies (especially when interrogating new technologies) permeate the discussion throughout. Marshall McLuhan's ideas animate the discussion. After establishing the benefits from these conceptual starting points, the volume examines each of the FIPs: FIP 1 in chapter 3, FIPs 2 and 4 in chapter 4, FIP 3 in chapter 5, and FIP 5 in chapter 6. Chapter 7 presents multi-layered recommendations.

A word of caution, even at this early phase: There are no easy solutions and no solutions that promise to completely resolve the issues at hand. My recommendations are contestable; some may not be feasibly or adequately adaptive. However, wringing our hands with worry is not enough, nor is throwing our hands up and running away screaming. The fact that people have not listened or taken action before is not proof that they are unable or unwilling to listen and change this time, or the next.

On the one hand, the data marketplace is so lucrative and so embedded within the computational media infrastructure that, barring war or cataclysm, there's no going back to the noncommercial Internet (the pre–data collection era). On the other hand, the marketplace is *so* lucrative that moving back toward the golden mean—with the invisible hand of the government and some restraint by both producers and consumers—would still leave loads of profitable activities available for all. On the one hand, resistance is futile. On the other hand, there's hope and it ain't over until a great song has been sung at the end.

Chapter 7 presents recommendations for actions that can be taken by all parties with skin in this game: our legislators and the government, judges and courts, large commercial entities that participate in the data marketplace, media and software developers and distributors, local businesspeople with digital aspects to their enterprise, and everyday citizen-users. These inscrutable and intractable problems cannot be solved by any single entity or by the actions of one group of participants. The Internet, digital and computational new media, and the data marketplace are complex systems requiring holistic and concerted efforts toward improvements. It is hoped that this volume contributes to the development of improved data management procedures that return commonsense approaches to privacy and surveillance in the United States.

· 1 ·

THE HARMS

This chapter documents numerous incidents of damages caused by improprieties in the data marketplace, focusing on three general areas of concern.

First, extensive damage is done when individuals lose control over personal data, especially when that loss of control results from or leads to lawlessness. Hacking and inappropriate data leaks expose victims to malicious harms, economic losses, and extensive drains on time, effort, confidence, and trust.

Second, misappropriation and mismanagement of private information can bring about circumstances that compromise constitutional principles. Both data breaches and intrusive data collection can become tantamount to self-incrimination, loss of due process, and violations of the guarantees to anonymous gatherings and communication (freedom of speech).

Third, the data marketplace thrives on the unequal value equation related to personal data. Industries using personal data to produce many billions of dollars in annual profits operate under the fictions that each piece of personal datum has little or no value and that value exists only after and as a result of aggregating data into sets for analysis. Nothing could be further from the truth. Individualized data has value but the marketplace (and our legislatures, our courts, and commercial enterprises) find it convenient (and profitable) to uphold the myths that deny individuals any profits from giving up their personal data. The myth of 'free' services also factors into the value equation.

Loss of Control over Personal Data

This volume examines two varieties of loss of control over personal data. In the first, lawless hackers maliciously steal data from commercial, nonprofit, or governmental data systems. Despite the fact that these systems are illegally hacked, these instances illustrate that organizations collecting and managing private data often fail to adequately protect it, thereby leaving personal information open to misappropriation. In the second variety, generally law-abiding entities either purposely or inadvertently engage in (or allow) inappropriate data collection or actually release information that should be protected.

Hacking as Lawlessness

The terms "computer hacking" and "hacker" have various definitions. As part of his Virginia Tech course "Hackers and Hacking," Michael J. McPhail provides definitions from two experts in early computing systems: Donn Parker differentiates among benign, unsavory, and malicious hackers, while William Landreth "describes five categories of crackers, each with a different motivation: the Novice, the Student, the Tourist, the Crasher, and the Thief" (McPhail, "Donn Parker's Categories of Hacker" 2016). Without discounting the positive or neutral uses of the term, this chapter examines outcomes of unlawful hacking as they relate to loss of privacy across the data marketplace.

Compiling a list of events that compromised large collections of personal data threatens to boggle the mind. The website *Hacked* labels 2015 "The Year of the Breach," citing data from the Identity Theft Research Center indicating that over 200 million personal records were exposed that year (Maras 2015). As of December 29 of that year, 780 breaches had exposed 177,866,236 records of personal data during 2015 (*ITRC Data Breach Report* 2015). Of those data breaches, 68% (121,629,812) occurred in the Medical/Health Care sector, 19.2% (34,222,763) in Government/Military, 9.1% (16,191,017) in the general category of Business, 2.8% (5,063,044) in Banking/Credit/Financial, and 0.4% (759,600) in Education.

Maras (2015) refers to a review by the public relations firm 10 Fold (*ITRC Data Breach Report* 2015) in which the following seven breaches are identified as the year's largest:

> Anthem
> The Anthem breach of 78.8 million patient records in early 2015 marked the largest breach in history. ... [T]he breach impacted an additional 8.8 to 18.8 million

non-patient records, including names, Social Security numbers, birth dates, employment data and addresses. ...

Excellus BlueCross Blue Shield
The breach compromised personal information of more than 10 million members and leaves members vulnerable to identity theft and fraud. ...

Premera Blue Cross
The health insurer discovered the attack affecting 11 million members in January of this year after it began in May of 2014. ...

VTech
VTech, the maker of tablets and gadgets for children, had kids' and parents' information compromised by the breach of the Kid Connect and Learning Lodge app store customer database. The breach affected 6.4 million kids and 4.9 million parent accounts globally and marked the first attack to directly target children. ...

Experian/T-Mobile
Attackers breached a server in a North American Experian/T-Mobile business unit containing personal ID information of about 15 million T-Mobile customers. ...

Office of Personnel Management
The attack affected 19.7 million individuals who applied for security clearances, plus 1.8 million relatives and other government personnel associates and 3.6 million former and current employees. The compromised data included 5.6 million fingerprint records that belong to background check applicants. ...

Ashley Madison
A hacker group called The Impact Team accessed the website's user database, including financial and proprietary information of 37 million users. (Maras 2015)

Of course, many other significant hacking-based data breaches occurred in 2015. In February, for example, the US government received "a massive cache of leaked data that revealed how the Swiss banking arm of HSBC, the world's second-largest bank, helped wealthy customers conceal billions of dollars of assets" (Lewis 2015). Data from the leaks were collected, collated, and turned over to the government by numerous legitimate news organizations, constituting "one of the biggest banking leaks in history, shedding light on some 30,000 accounts holding almost $120bn (£78bn) of assets. Of those, around 2,900 clients were connected to the US, providing the IRS with a trail of evidence of potential American taxpayers who may have been hiding assets in Geneva" (Lewis 2015).

In April 2015, WikiLeaks published a searchable database of every email and document that had been stolen from Sony, via hacks, earlier in the year

(Pagliery 2015). In an introductory statement, WikiLeaks said the archive "contained 30,287 documents, along with 173,132 emails from more than 2,200 addresses" (Cieply 2015).

The proliferation of malicious hacking is illustrated in stunning profundity by repeated data losses at the Internet Corporation for Assigned Names and Numbers (ICANN), the organization responsible for allocating Internet domain names and IP addresses. In a July 2015 attack, hackers "gained access to usernames, email addresses, and encrypted passwords for profile accounts on ICANN.org public website" (Khandelwal 2015). This attack followed an earlier one, in November 2014, when a hacker gained access to "ICANN systems, including the Centralized Zone Data System (CZDS), the domain registration Whois portal, the wiki pages of the ICANN Governmental Advisory Committee (GAC), and ICANN blog" by tricking (spear-phishing) ICANN employees—a generally pretty sophisticated group of computer users—into handing over their online credentials (log-ons/passwords) (Khandelwal 2015).

Further evidence of the rapid spread of criminal hacking is that it is now the leading cause of data breaches across healthcare industries, where "criminal attacks … are up 125 percent compared to five years ago. Now, in fact, nearly 45 percent of data breaches in healthcare are a result of criminal activity" ("Criminal Attacks" 2015).

A 2016 report by the private intelligence firm Flashpoint notes the substantial increase in internationally based cybercrime focused on the data of US citizens—on individuals rather than governments or large organizations:

> [F]or all the dangers of state-backed, politically motivated hacking, profit-motivated cybercriminals are a more frequent and perhaps much greater hazard for the majority of people. … [I]ndividuals are much softer targets than governments or major corporations. And every individual has access to information—ranging from medical data to bank-account numbers to online passwords to basic biographical information—off which enterprising hackers can profit. (Rosen 2016)

Although purposive and criminal hacking proliferates, and is on the rise, somewhat more benign forms of data mismanagement also strongly compromise personal data privacy.

Leaks and Sponges: We Were Not Supposed to Collect It or Lose It

Prolific amounts of private information are mismanaged when organizations, private and public, release or otherwise lose control over significant numbers of records. In these instances, no malicious hacking is involved. Rather, the organization simply fumbles its role as information gatekeeper. Sometimes data are lost as the result of something as simple as an employee losing a laptop (Vijayan 2011).

Often, private information is revealed by organizations that have collected it legally and for valid purposes only to see their enormous databases compromised via mistakes or technical snafus. For example, in December 2015,

> Information about 191 million registered voters was left exposed on the Internet, raising new questions about the security chops of political campaigns who increasingly hold large caches of data about Americans. The leak appeared to be the result of a technical error that allowed the information to be publicly accessed online, not a hack. ... The leaked data contains a slew of information about Americans who are registered to vote, including their full name, home address, mailing address, date of birth, phone number, political affiliation, emails and details about if they had voted in each election going back to 2000. Much of this information is public record and a variety of companies sell political campaigns databases that combine the information so they can target their outreach. But some states have different restrictions on how the information may be disclosed or how it can be used. (Peterson, "Massive Database" 2015)

Publishers of the report stated that "it remains unclear who controls the database" (Peterson, "Massive Database" 2015). You read that right: 15 years' worth of records about 191 million registered voters made their way to the public Internet, and security researchers weren't even able to identify the owner of the database from which the records were released.

Even companies with software and systems purported to protect users' security are subject to lax or dishonest data management practices. The FTC fined LifeLock twice—in 2010 ($12M) and in 2015 ($100M)—for deceptive advertising and false claims about the range and effectiveness of their privacy protection software and systems ("LifeLock to Pay" 2015; "LifeLock Will Pay" 2016).

While the LifeLock judgments were fines for bad-faith promotions, other data protection outfits have suffered losses of private data with real costs to consumers. In December 2015, security researcher Chris Vickery "found and reported a massive security issue on the Web servers of MacKeeper ... [where] the databases of Kromtech, the company behind MacKeeper, were open to external connections and required no authentication whatsoever. The names, passwords, and other information of around 13 million users may have been exposed" in a breach that Kromtech admitted (Degeler 2015).

Situated between malicious hackers and accidental data dumpers are organizations using technologies that collect data surreptitiously, thereby exposing users to mismanagement risks they didn't even know they were taking. Although many computer applications and processes collect user data without consumers' conscious awareness, some practices are so secret as to violate even a permissive person's sense of propriety.

Although Apple maintains that apps approved for inclusion in its App Store do not collect data about users, numerous violations of that policy have occurred. In fact, researchers have found "more than 250 iOS apps that violate Apple's App Store privacy policy" (Goodin, "Researchers Find" 2015):

> The apps, which at most recent count totaled 256, are significant because they expose a lapse in Apple's vetting process for admitting titles into its highly curated App Store. They also represent an invasion of privacy to the one million people estimated to have downloaded the apps. The data gathering is so surreptitious that even the individual developers of the affected apps are unlikely to know about it, since the personal information is sent only to the creator of the software development kit used to deliver ads. (Goodin, "Researchers Find" 2015)

Finding malicious software that collects data from unsuspecting users, in violation of Apple App Store policies, is not a one-time occurrence. Although Apple is vigilant, malicious apps find their way into circulation despite the company's vetting mechanisms.

> Apple officials are cleaning up the company's App Store after a security firm reported that almost 40 iOS apps contained malicious code that made iPhones and iPads part of a botnet that stole potentially sensitive user information. The 39 affected apps—which included version 6.2.5 of the popular WeChat for iOS, CamScanner, and Chinese versions of Angry Birds 2—may have been downloaded by hundreds of millions of iPhone and iPad users, security researchers said. (Goodin, "Apple Scrambles" 2015)

Clearly, Apple is not the only technology outfit with issues of this nature. Uber, for instance, "has acknowledged that a flaw in its software caused it to leak personal data belonging to … about 700 of its 'partners' [drivers] in the U.S. … [The leak] exposed … social security numbers, photos of driver licences, tax forms and other details" ("Uber Error" 2015). And, of course, Target—though no stranger to the high costs of data breaches, recently offering up to $10M to settle a class-action suit over its 2013 breach (Halzack 2015)—remains unable or unwilling to control its data collection and management procedures. "According to researchers from security firm Avast, the database storing the names, e-mail addresses, home addresses, phone numbers, and wish lists of Target customers is available to anyone who figures out the app's publicly available programming interface" (Goodin, "Wish List App" 2015).

Costs and Harms of Data Breaches

Calculating consumers' costs and harms from data breaches is a daunting task. Many of the damages suffered by consumers are indirect losses of time, effort, and trust in the system; other costs result from secondary breach effects. While some consumers spend years mopping up the damages from, for example, identity theft, most consumers involved in data breaches suffer very small direct monetary damages. For example, in a study of 16.6 million people affected by identity theft in 2012, "of the victims who experienced an out-of-pocket loss, about half lost $99 or less" (Bureau of Justice Statistics 2013).

Cost and loss analyses are not available across a wide range of privacy breaches, so data about identity theft are especially revealing. While 66% of identity theft victims suffer some financial loss, and the overall total of that loss reaches almost $10,000, only 14% of identity theft victims pay these costs out of pocket; most pay less than $100 (DiGangi 2013). Time and effort costs can be higher. For example, 10% spend over a month resolving financial and credit problems ("How Consumers Foot the Bill" 2014). Another analysis—of breach costs in 2013 in the state of Utah—estimated that victims "will spend about 20 hours and $770 on lawyers and time lost from work to resolve the case" (Carrns 2013).

As noted, most costs and harms to consumers are indirect. The really extensive costs to consumers result from increased prices across marketplaces racked with enormous financial losses from breaches of private data (whether malicious or unintended). The Ponemon Institute reports that "the average total cost of a data breach for the participating companies increased 23 percent over the past two years to $3.79 million. ... The average cost paid for each lost or stolen record containing sensitive and confidential information increased 6 percent, jumping from $145 in 2014 to $154 in 2015" (Ponemon 2015; see Ponemon Institute 2015).

Security systems provider Authentify describes how those indirect costs are passed on to consumers:

> Retailers pass these costs on through higher prices.
> - Where retailers are on the hook for fraudulent online orders, their insurance pays the bill ..., which is passed on to shoppers through higher prices.

- Credit providers and banks have to roll back fraudulent charges made on stolen cards ... and have to reissue cards for anyone who might be affected.
- These costs are passed on to consumers through higher interest rates and credit card fees, up to a 5–7% increase in some cases. ("How Consumers Foot the Bill" 2014)

As I will show in chapter 3, the secrecy that surrounds the collection of private data violates the first FIP. Distribution of the costs of data breaches is done in secret, in plain sight. Consumers have no way to perceive, understand, or appreciate price increases that are based on organizations recovering financial losses due to data breaches. Because annual losses stretch into the billions of dollars, one has to believe that consumers are paying significantly increased overall prices as a result of the trust they place in the organizations that hold and use private data.

Undermining Constitutional Protections

Numerous authors have long discussed challenges to constitutional privacy protections brought about by new media technologies. Far from being alone, Solove (2004, 2011) presents evidence of the declining strength of traditional legal interpretations concerning privacy. The current section addresses shortcomings across five constitutional amendments (the First, Third, Fourth, Fifth, and Fourteenth) and within Article III.

Encroaching on First Amendment Protections

The First Amendment promotes free speech, protects citizens' rights to peaceably assemble, and constrains the government from prior restraint. Arguments and litigation over the government's massive data collection programs often hinge on the damage (real or potential) to personal privacy, and liberty, that such programs render. Opponents of the programs, advocating civil liberties, argue that excessive, intrusive, surreptitious surveillance undermines citizens' abilities to communicate, freely and without fear of reprisals.

Prior to the age of information, constitutional law provided what generally seemed to be adequate protections of citizens' privacy rights. However, American systems of justice have not yet embraced cyberspace as a "place" (see *Estavillo v. Sony Computer Entertainment America* 2009), and so generally do not extend First Amendment protections against online and/or digital data collection and surveillance. Often, the protections that citizens think are, or should be, in place are not.

In the early days of new media, users could exchange electronic communication anonymously or using pseudonyms; often their interactions with websites left little data for individualized analysis. These features encouraged early new media enthusiasts to opine (and to hope) that new media would have wildly democratizing effects on society. Freedom of speech looked to be a promising outcome of new media communications.

As computational electronically mediated communication matured, the tracking of users' online and mobile interactions became very sophisticated and exceedingly personalized. So many data points are now available (and collected) that precious little information and few interactions are needed to identify individuals. In fact,

> it is possible to assign location to WiFi access points based on a very small number of GPS samples and then use these access points as location beacons. Using just one GPS observation per day per person allows us to estimate the location of, and subsequently use, WiFi access points to account for 80% of mobility across a population. These results reveal a great opportunity for using ubiquitous WiFi routers for high-resolution outdoor positioning, but also significant privacy implications of such side-channel location tracking. (Sapiezynski et al. 2015)

Participants in the data marketplace collect and utilize users' private data, generally without their full knowledge or their conscious, willful, consent. The collection and leveraging of users' individual (and generally private) information impinges on their personal freedom. Generally, however, appeals to First Amendment protections neither dissuade the government and/or commercial interests from leveraging personal data nor serve as effective litigation tools for users.

For example, the Electronic Privacy Information Center (EPIC) asked the FTC to sanction Uber for collecting location data even when their app is running in the background. Uber's so-called privacy policy is typical in that instead of protecting a user's privacy, it notifies the user of the various ways that Uber will violate their privacy. The policy reads, in part, as follows:

> When you use the Services for transportation or delivery, we collect precise location data about the trip from the Uber app used by the Driver. If you permit the Uber app to access location services through the permission system used by your mobile operating system ("platform"), we may also collect the precise location of your device when the app is running in the foreground or background. We may also derive your approximate location from your IP address. (quoted in Farivar, "FTC Asked" 2015)

In January 2016, Uber came to a settlement agreement with the Attorney General of the State of New York over its data collection methods and failures to notify authorities about data breaches. Uber agreed to amend its policies, institute new practices, and pay a $20,000 fine for inadequate reporting ("A.G. Schneiderman" 2016).

Knowing a user's location enables interested parties, be they commercial or governmental, to identify where the user lives, works, and plays as well as to draw (often accurate) inferences about their activities. Catherine Crump, a law professor at the University of California, Berkeley, illustrates: "Do they park regularly outside the Lighthouse Mosque during times of worship? They're probably Muslim. Can a car be found outside Beer Revolution a great number of times? May be a craft beer enthusiast—although possibly with a drinking problem" (quoted in Farivar, "We Know Where" 2015).

Government, especially law enforcement, has enthusiastically embraced location technologies for use in ways that compromise privacy rights associated with free assembly and freedom from prior restraint. For example, city governments have used data from license plate readers to threaten and shame citizens who drive in 'suspect' areas of town, without a stick of proof that the person is there for illegal purposes. In Los Angeles, the "city council proposed using license plate readers to identify vehicles driving in areas known for prostitution and sending 'john letters' to the registered owner's address" (Storm 2015). In 2012, cities in as many as 40 states sponsored programs that involved sending "Dear John" letters to suspected "johns," connecting hand-collected license data with mail addresses.

The City of Oakland collected more than 4.6 million reads of over 1.1 million unique plates between December 23, 2010, and May 31, 2014. The use of license plate readers to identify vehicles in particular locations fails to account for those drivers (and their passengers) who are innocently (even mistakenly) in, or passing through, those locations. This issue will be mentioned again in the later section about the erosion of Fifth Amendment protections from self-incrimination.

Further, data of this type facilitate the identification of particular aspects of individual lives with the potential of chilling free association. The online technology publication *ars technica*, in a 2015 journalistic investigation using the publicly accessible Oakland data, clearly demonstrated the revelatory potential of that data:

> Anyone in possession of enough data can often—but not always—make educated guesses about a target's home or workplace, particularly when someone's movements

are consistent (as with a regular commute). For instance, during a meeting with an Oakland city council member, Ars was able to accurately guess the block where the council member lives after less than a minute of research using his license plate data. Similarly, while "working" at an Oakland bar mere blocks from Oakland police headquarters, we ran a plate from a car parked in the bar's driveway through our tool. The plate had been read 48 times over two years in two small clusters: one near the bar and a much larger cluster 24 blocks north in a residential area—likely the driver's home. (Farivar, "We Know Where" 2015)

It has long been evident that Internet Service Providers (ISPs) hold the key to an enormous treasure trove of user data. As these large entities gain market share, they are under increased scrutiny by and pressure from privacy advocates, as little (or no) legislation constrains their use of private user data for commercial purposes. According to Chris Hoofnagle, a law professor at the University of California, Berkeley, "An ISP has access to your full pipe and can see everything you do" online if you aren't taking extra steps to shield your activities. Further,

> Other than corporate privacy policies, he said, nothing under current law prevents broadband companies from sharing information with marketers about what types of Web sites you visit. Broadband companies have shown an increasing interest in consumer data. Verizon last year bought AOL in a $4 billion deal that allowed the telecom firm to integrate AOL's substantial advertising technology into its own business, improving how it targets ads on Internet videos. (Fung, "Internet Providers" 2016)

Along similar lines, AT&T uses pricing power to leverage data collection targeted at their gigabit service customers. The company is "boosting profits by rerouting all your Web browsing to an in-house traffic scanning platform, analyzing your Internet habits, then using the results to deliver personalized ads to the websites you visit, e-mail to your inbox, and junk mail to your front door" (Brodkin, "AT&T's Plan" 2015).

Users have been generally unsuccessful with legal challenges focusing on data marketplace use of personal information. For instance, a woman in the state of New York brought a civil suit against AMC Networks Inc. for aiding Facebook's tracking of her viewing and online behaviors and habits. In the suit, the 62-year-old argued that "Facebook is uniquely able to directly link the data they accumulate on individuals' digital behaviors with the additional personal data that it extracts from its users' Facebook accounts," and that, "[w]hen combined, this data reveals deeply personal information about a consumer." The US district judge ruled against the plaintiff's claims. But rather than examine the question of whether the data marketplace actions were legal (or ethical), the judge focused on whether the user qualified, under AMC's terms of service, for legal judgments based on those terms. The judge ruled that since the woman had not formally subscribed to AMC's services, she was not entitled to relief regarding their various functions (Gershman, "Judge: Browsing Data" 2015).

In short, constitutional protections of free speech, of free assembly, and against prior constraint simply do not stretch to protect individuals' data from data marketplace abuses.

Apparently, traditional constitutional protections do not always protect corporate entities in the new media environment either. At various times, many of the major social media and telecommunication companies have made an argument similar to the one proposed by Twitter, that the government's "virtual ban of detailing the scope of US surveillance on the microblogging site is an unconstitutional 'prior restraint' of speech protected by the First Amendment" (Kravets, "Twitter Says Gag" 2014).

The history of the back-and-forth battles of new media companies and the telecoms against the federal intelligence community is replete with evidence that traditional constitutional protections of the First Amendment might no longer apply in the new media environment. On the one hand (and going back to just after enactment of the Patriot Act in 2001), courts have (sometimes) found that various aspects of the mass data collection practices of the NSA violate constitutional principles. For example, a federal district court judge in San Francisco ruled in 2013

> that National Security Letter (NSL) provisions in federal law violate the Constitution. The decision came in a lawsuit challenging a NSL on behalf of an unnamed telecommunications company represented by the Electronic Frontier Foundation (EFF). ... Judge Susan Illston ordered that the Federal Bureau of Investigation (FBI) stop issuing NSLs and cease enforcing the gag provision in this or any other case. ... [T]he court held that the gag order provisions of the statute violate the First Amendment. ("National Security Letters" 2013)

Judge Illston ordered the FBI to stop issuing unconstitutional "spying orders"—a.k.a. NSLs—and to stop enforcing the gag orders that go along with them.

Less than a year later, however, the very same judge upheld the gag order procedures that she had earlier found to be unconstitutional. Despite Judge Illston's earlier order, the government continues to use the gag orders and media and telecom companies continue to bring litigation against the practice. In the 2014 case, Judge Illston reasoned that since her ruling was under appeal at the Ninth Circuit she was legally compelled to treat the gag order protocols that were in place prior to her ruling as the status quo (Zetter 2014).

Social media and telecom companies continue to bring legal action against the practice in support of their First Amendment rights. In late 2014, Twitter sued over the prior restraint issue on First Amendment grounds (Kravets, "Twitter Says Gag").

Hopes that the appeal of the Illston ruling would settle the matter were dashed in 2015 when the Ninth Circuit of Appeals sent the case back to the district courts over technicalities (Cohn and Opsahl 2015).

In his advocacy of privacy rights in the face of new technologies, Louis Brandeis—initially in coauthorship with Samuel D. Warren and both before and after becoming Chief Justice of the Supreme Court—maintained the right to be left alone (Warren and Brandeis 1890). Later, in 1928's *Olmstead v. United States*, Brandeis presciently expressed the opposition that he would likely have against today's electronic data collection practices: "Discovery and invention have made it possible for the Government, by means far more effective than stretching upon the rack, to obtain disclosure in court of what is whispered in the closet" (*Olmstead* 1928). Current data marketplace practices encroach on First Amendment protections.

Encroaching on Third Amendment Protections

The Third Amendment places restrictions on the quartering of soldiers in private homes without the owners' consent, prohibiting it during peacetime. In a 2015 opinion column in the *Los Angeles Times*, Mike Gatto—lawyer, assemblyman from California's 43rd (L.A. County) District, and chairman of the Assembly's Consumer Protection and Privacy Committee—argues that the Third Amendment should be put to use in protection against privacy invasions wrought by intrusive electronic data collection:

> The British could spy on American colonists by keeping soldiers among them. Today, the government can simply read your email. Centuries ago, patriots wrote angry letters about soldiers observing the ladies of the house at various stages of undress. Now ... the NSA can just view your intimate selfies. If you're like me and you text your wife to pick up a gallon of milk, your smartphone is an integral part of internal household communication. ... [W]hen your refrigerator, your thermostat, your sprinklers and your TV connect to the Internet and gather data on your household habits, can it really be doubted that an intrusion into these systems is an intrusion into your home? ... [T]he military has lodged itself in your home when it invades your personal space on your devices. Indeed, if the 3rd Amendment only applied to real estate, it would

be moot, since the 5th Amendment provides a mechanism for compensation when government uses property even briefly. I'm not alone in seeing the ever-expanding federal government's military-run surveillance as a modern form of quartering troops in our homes. Several mainstream but inventive constitutional law professors have argued that the 3rd Amendment applies to surveillance. Like me, they see that the ubiquitous incorporeal presence of a military agency in our household systems is as significant to us as the physical presence of redcoats was in the 1700s. Let's dust off the 3rd Amendment to make this point, and soon. (Gatto 2015)

Gatto's concerns about government data collection—analogous, in his argument, to hosting a spy in one's home, and thereby violating Third Amendment protections—do not apply directly to commercial operations. Nevertheless, our homes are evermore filled with devices that spy on us. In some cases, the commercial enterprises merely report data that can be used for targeted marketing. As noted earlier, however, a significant overlap between commercial and governmental activities occurs when data streams cross in the data marketplace.

Andrew Whalen reports on a newfangled version of a toy commonly found in the home, the Hello Barbie doll, that "contains a Wi-Fi connected microphone that will send snippets of conversation to an external server and reply with 'more than 8,000 lines of recorded content'" (Whalen 2015). Whalen raises questions about relationships among funders, developers, and governmental interests in the toy. These connections highlight the nebulous and sometimes questionable relationships between new media enterprises and the government and remind us that with the advent of the Internet of Things/Everything, an increasing number of devices will collect data within the home.

> ToyTalk, the company behind the conversation software that records Hello Barbie chatter and designs her responses ..., offers a robust privacy policy, with some notable loopholes. ... ToyTalk will use Hello Barbie recordings to "develop, test or improve speech recognition technology and artificial intelligence algorithms." ToyTalk also can send recorded data to "other service providers," offering Microsoft as an example ... [as well as] intelligence agencies: what they call "lawful requests or legal authorities" with "lawful subpoenas, warrants, or court orders." It would be paranoid to imagine the CIA or NSA investing money in a children's software company with the specific intent of gathering intelligence or building psychological profiles. ... However, it starts sounding a little less paranoid when looking at ToyTalk's investors. Howard Cox is a "Special Limited Partner" with Greylock Partners, one of the "seed round" venture capital funders for ToyTalk. Howard Cox is also on the board of the Brookings Institution and In-Q-Tel. Both organizations are hip deep in CIA connections. Khosla Ventures, another major investor in ToyTalk, invested in database company MemSQL with In-Q-Tel. So did ToyTalk investor First Round Capital. It remains

unlikely that the CIA or NSA has a strong desire to hoover up the ramblings of little girls, but there's just enough smoke to justify a modicum of paranoia surrounding Hello Barbie Doll. (Whalen 2015)

Data collection and analysis by government, law enforcement, and the intelligence community has been enhanced by the ubiquity of information provided via commercial enterprises. One supposes that few social media or telecom outfits set up data collection protocols for the express purpose of spying on American citizens at the behest of the government. However, commercial enterprises have been virtually helpless in stopping government entities from demanding that they share the data they have collected. Ubiquitous data collection devices in the home compromise Third Amendment protections of privacy.

Encroaching on Fourth Amendment Protections

The Fourth Amendment protects against unreasonable searches and seizures by setting requirements for search warrants based on probable cause. Significant battles over data from phones have been at the center of disagreements between the government and privacy advocates. This is perhaps the lone aspect over which 'the people' appear to be holding their own against the government and new media industries that provide private data (often unwillingly) to the government.

In June of 2014, the Supreme Court ruled in *Riley v. California* (joined with *United States v. Wurie*) that "[t]he police generally may not, without a warrant, search digital information on a cellphone seized from an individual who has been arrested" (*Riley* 2014). Writing for *SCOTUSblog*, Marc Rotenberg and Alan Butler (2014) noted, "The Court's conclusion ... will affect not only digital search cases, but also the NSA's bulk record collection program, access to cloud-based data, and the third-party doctrine."

> The Court rejected outright the government's proposal that agencies "'develop protocols to address' concerns raised by cloud computing." The Chief Justice stated plainly that "the Founders did not fight a revolution to gain the right to government agency protocols." ... [I]n *Riley* the Court also explicitly rejected the government's argument that call logs and other "metadata" are not deserving of Fourth Amendment protection. The Court's argument takes clear aim at the third-party rule—that "non-content" records like call logs, location data, and other metadata held by third parties can be collected by the government without a warrant. Like the data stored on cell phones, metadata can reveal "an individual's private interests and concerns

... can also reveal where a person has been" and there is an "element of pervasiveness" in the collection of all metadata records about an individual. (Rotenberg and Butler 2014)

This ruling implies that law enforcement officials should get a warrant before leveraging cell phone records as part of ongoing investigations. However, numerous instances verify that, sometimes, getting law enforcement operatives in the US to comply with the spirit or letter of constitutional protections for privacy is a lot like playing a game of Whac-A-Mole: No sooner does one type of data collection require proper legal certification than law enforcement starts to depend on a different technological solution to many of the same law enforcement challenges, in ways that evade or eviscerate constitutional protections.

The American Civil Liberties Union, among other organizations, has complained mightily about the use by police nationwide of 'stingray' devices that enable the collection of data related to cell phone calls, including the number of the phone making a call, the number of the phone being called, and the exact location of the caller (Saavedra 2015). Stingrays can also be used to intercept calls and text messages by 'spoofing' cell towers through which mobile phone traffic is routed. In San Bernardino County (California's fifth-largest county and the twelfth-most-populous county in the country), the sheriff "has deployed a stingray hundreds of times without a warrant, and under questionable judicial authority" (Farivar, "County Sheriff" 2015).

Spotty-to-non-existent compliance with warrant requirements is rampant in the use of stingray devices. For example, when law enforcement agencies request purchases of equipment through the FBI, they are forced to sign a nondisclosure agreement that binds them to secrecy in the use of those devices (Farivar, "Sheriff" 2015). Although the Department of Justice has instructed agencies under its control to seek a warrant before using stingrays, in many jurisdictions no such requirement exists. For example, in Sacramento County, California, the sheriff's department has decided to seek what it refers to as "judicial approval" rather than a warrant (Farivar, "County Sheriff" 2015). The legal status of the so-called judicial approval is questionable; in contrast, the status of acquiring a warrant based on probable cause is clear as a bell.

In short, data collection that compromises the Fourth Amendment's protection against illicit searches and seizures is deeply ingrained in government activities, especially those of the law enforcement and intelligence communities.

Encroaching on Fifth Amendment Protections

The Fifth Amendment to the Constitution protects American citizens against requirements that they self-incriminate. The various tracking functionalities discussed with regard to First Amendment protections also have Fifth Amendment implications. In particular, license plate reader systems and/or stingray-based cell phone interceptions run the risk of forcing citizens to unwittingly and unwillingly provide evidence against themselves.

Most smart gadgets record data that can be used against their users. Increasingly, courts hearing cases involving technology and privacy are faced with decisions that have constitutional implications under the Fifth Amendment. In *Arias v. Intermex Wire Transfer* (2015), for example, "Myrna Arias, a former Bakersfield sales executive for money transfer service Intermex, claims ... that her boss ... fired her shortly after she uninstalled the job-management Xora app that she and her colleagues were required to use ... [and that] allowed her ... 'bosses to see every move the employees made throughout the day'" (Kravets, "Worker Fired" 2015). Arias' suit claimed invasion of privacy, retaliation, and unfair business practices, seeking damages in excess of $500,000.

Networked new media devices raise the stakes regarding the potential for self-incriminating behavior. As devices produce, record, and report data without the conscious and willing consent of the people producing the information (the users of the devices), increasing numbers of citizens will find themselves facing commercial and legal difficulties over actions that would, in the past, either have gone unnoticed or been reported by participants in their own ways. For example, there are significant and realistic concerns that smart cars will produce data that might harm their operators, by returning reports to insurance companies. Is not uncommon for even diligent drivers to occasionally ignore or exceed a driving parameter established by law. Prior to smart cars, such information only reached insurance companies when negative events resulted in accidents or citations. Privacy advocates and some citizens are concerned that new technologies will return much more significant amounts of data to insurance companies, leading to rate increases or other changes in policy terms. A recent incident raises these kinds of questions and concerns:

> A Florida woman was reportedly arrested and placed into custody last week, after her car implicated her in at least one alleged hit-and-run incident. ... [A] car driven by 57-year-old Cathy Bernstein automatically called 911 to report a crash. The call was part of a safety feature designed to help first responders locate people who may

have lost consciousness in crashes. ... The vehicle was in a collision, and called 911 through the driver's phone, which was paired with the car. When the driver did not respond to the operator, the car appears to have taken over and provided the operator with the information needed to locate the vehicle. (Tsukayama 2015)

A number of aspects of this incident stand out. First, this particular incident involved lawless behavior: a hit-and-run accident and driver. In this specific instance, perhaps even privacy advocates would admit that the society as a whole benefits from bringing this kind of lawlessness under control and that technology that helps to do so is to be applauded, not denigrated and constrained. Second, one can imagine circumstances only slightly different leading to similar positive conclusions; had the driver been disabled by the accident, and perhaps unable to answer the call due to injury, the ability of rescue responders to quickly locate the vehicle might have been a lifesaver.

However, it does not take a vivid imagination to conjure up circumstances in which the outcomes would strongly compromise Fifth Amendment protections against self-incrimination. For example, people who are involved in divorce proceedings sometimes visit locations that, if identified, might be used against them in settlement proceedings. Likewise, during investigations of criminal activity, perfectly innocent subjects can fall under scrutiny and overbearing examinations by law enforcement officials merely because of their presence at or near the scene of a crime. Even if their presence is innocent, law enforcement officials can refuse to exclude them from consideration. Such individuals might not have volunteered information about their location/presence, nor even known they were at or near the scene of a crime. Nevertheless, automated reports sent by smart gadgets produce information that compromises Fifth Amendment protections against self-incrimination.

Encroaching on Fourteenth Amendment Protections

The Fourteenth Amendment bars a state from depriving a person of liberty without due process of law. This includes protecting the fundamental right to privacy. As is the case with other constitutionally based protections, strict interpretations limit the protections to constraints against the government. However, this approach clouds the fact that privacy violations by commercial enterprises often support governmental actions. Since the data streams of commercialism and government are now thoroughly intertwined in the data marketplace, claiming that a given activity is only commercial often fails the smell test.

The controversial landmark case *Roe v. Wade* established abortion rights/rights for choice based on privacy protections tied to the Fourteenth Amendment. There is, then, irony in the contemporary information-gathering practices of antiabortion activists. In the recent past, "abortion opponents have become experts at accessing public records such as recordings of 911 calls, autopsy reports, and documents from state health departments and medical boards, then publishing the information on their Web sites" (Ornstein, "Abortion Foes" 2015). However, these practices have now been 'upgraded' to waiting "outside clinics, [and] tracking or taking photos of patients' and staffers' license plates and ambulances."

Medical information is often exceedingly private from the point of view of the person represented by the data. Breaches of this type of private information are likely to raise grave concerns on the part of the person providing the data. For example, in 2014 an Alaska citizen filed litigation in state district court referencing Alaska's Genetic Privacy Act. Michael Cole became the lead plaintiff in a class-action suit—*Cole v. Gene by Gene, Ltd.* (2014)—because, "after purchasing a Family Tree at-home genetics kit and joining a 'project,' an online forum for people doing related research about their ancestors, 'the results of his DNA tests were made publicly available on the Internet, and his sensitive information (including his full name, personal e-mail address, and unique DNA kit number) was also disclosed to third-party ancestry company RootsWeb'" (Farivar, "Lawsuit" 2014).

Perhaps the most stunning challenges to privacy rights protected by the Fourteenth Amendment are brought about by the ever-burgeoning use of facial recognition technologies. Facial recognition is now a ubiquitous part of both commercial and governmental data marketplace procedures. Commercial operations like Facebook use facial recognition for a wide variety of purposes related to targeted marketing. Law enforcement and the intelligence community use facial recognition in efforts to identify and track criminals and criminality. Additionally, the technology has been used "in retailer anti-fraud programs, for in-store analytics to determine an individual's age range and gender to deliver targeted advertising, to assess viewers' engagement in a videogame or movie or interest in a retail store display, to facilitate online images searches, and to develop virtual eyeglass fitting or cosmetics tools" (Neuburger, "Facial Recognition" 2015).

These efforts encroach on protections previously provided by both the fifth and fourteenth amendments, as facial recognition personalizes electronically collected information in truly individualized ways. Combining facial recognition with the myriad of other data collection techniques and protocols removes all senses of anonymity and/or aggregation from collected information.

There is little doubt that massive databases filled with facial imagery are being developed by law enforcement and intelligence. It is estimated that the FBI alone maintains a database with over 52 million images (Kravets, "FBI" 2014; Lynch 2014). Of particular concern to privacy advocates is the fact that the FBI is planning to include facial images of large numbers of non-criminals, with serious implications for the ways that electronic records in databases overlap:

> In the past, the FBI has never linked the criminal and non-criminal fingerprint databases. This has meant that any search of the criminal print database (such as to identify a suspect or a latent print at a crime scene) would not touch the non-criminal database. This will also change with NGI [the Next Generation Identification database]. Now, every record—whether criminal or non—will have a "Universal Control Number" (UCN), and every search will be run against all records in the database. (Lynch 2014)

Some states have moved to protect their citizens from inappropriate uses of biometric information technologies. For example:

> Texas and Illinois have existing biometric privacy statutes that may apply to the collection of facial templates for online photo tagging functions. Illinois's "Biometric Information Privacy Act," ("BIPA") 740 ILCS 14/1, enacted in 2008, provides, among other things, that a company cannot "collect, capture, purchase, receive through trade, or otherwise obtain a person's ... biometric information, unless it first: (1) informs the subject ... in writing that a biometric identifier or biometric information is being collected or stored; (2) informs the subject ... in writing of the specific purpose and length of term for which a biometric identifier or biometric information is being collected, stored, and used; and (3) receives a written release executed by the subject of the biometric identifier or biometric information." [740 ILCS 14/15(b)]. The Texas statute, Tex. Bus. & Com. Code Ann. §503.001(c), enacted in 2007, offers similar protections. (Neuburger, "Facial Recognition" 2015)

However, as Neuburger points out, "[n]either statute has been interpreted by a court with respect to modern facial recognition tools" ("Facial Recognition" 2015).

Improving facial recognition technologies has long been at the center of many corporate R&D missions. Technology belonging to and used by companies like Facebook and Google is now so advanced that it can identify people in pictures even when their faces don't show. "The day when companies like

Facebook and Google will be able to recognize you in pictures with 99% accuracy, even when your face doesn't show, is fast approaching. ... Facebook's DeepFace ... can tell whether the subjects in two different photographs are the same person with 97% accuracy" (Elgan 2015).

There is an area of constitutional privacy protection that appears to offer tools that privacy advocates and citizens can use to resist government-sponsored intrusions. Before discussing this use of Article III, I want to summarize a troublesome privacy scenario based on current practices.

In a January 10, 2016, article in the *Washington Post*, Fairfax County justice reporter Justin Jouvenal provides a chilling description of the ways that the Fresno, California, police use big data. Many other law enforcement entities and jurisdictions utilize similar systems. For example, "[p]lanes outfitted with cameras filmed protests and unrest in Baltimore and Ferguson, Mo. ... [D]ozens of departments used devices that can hoover up all cellphone data in an area without search warrants. Authorities in Oregon are facing a probe after using social media-monitoring software to keep tabs on Black Lives Matter hashtags" (Jouvenal 2016). Fresno's police department is willing to make its efforts somewhat more transparent than others.

According to Jouvenal, software used by police in Fresno is able to scour "billions of data points, including arrest reports, property records, commercial databases, deep Web searches and ... social media postings" and post the results. He continues:

> On 57 monitors that cover the walls of the center, operators zoomed and panned an array of roughly 200 police cameras perched across the city. They could dial up 800 more feeds from the city's schools and traffic cameras, and they soon hope to add 400 more streams from cameras worn on officers' bodies and from thousands from local businesses that have surveillance systems. ... Officers could trawl a private database that has recorded more than 2 billion scans of vehicle licenses plates and locations nationwide. If gunshots were fired, a system called ShotSpotter could triangulate the location using microphones strung around the city. Another program, called Media Sonar, crawled social media looking for illicit activity. Police used it to monitor individuals, threats to schools and hashtags related to gangs. (Jouvenal 2016)

Perhaps the most interesting aspect of the Fresno efforts is the department's beta testing of a threat scoring system software called Beware. And perhaps we should indeed beware:

> As officers respond to calls, Beware automatically runs the address. The searches return the names of residents and scans them against a range of publicly available

data to generate a color-coded threat level for each person or address: green, yellow or red. ... The Fresno City Council called a hearing on Beware in November after constituents raised concerns. Once council member referred to a local media report saying that a woman's threat level was elevated because she was tweeting about a card game titled "Rage," which could be a keyword in Beware's assessment of social media. Councilman Clinton J. Olivier ... said Beware was like something out of a dystopian science fiction novel and asked Dyer a simple question: "Could you run my threat level now?" Dyer agreed. The scan returned Olivier as a green, but his home came back as a yellow, possibly because of someone who previously lived at his address, a police official said. "Even though it's not me that's the yellow guy, your officers are going to treat whoever comes out of that house in his boxer shorts as the yellow guy," Olivier said. "That may not be fair to me." He added later: "[Beware] has failed right here with a council member as the example." (Jouvenal 2016)

Facial recognition, ShotSpotters, cell-tower-faking stingrays, license plate readers, "the FBI's $1 billion Next Generation Identification project, which is creating a trove of fingerprints, iris scans, [and] data from facial recognition software and other sources" (Jouvenal 2016), database-wide requests/demands for phone records from telecoms, citywide video surveillance (as one finds in urban areas such as Chicago)—the list of surveillance and data collection techniques appears to be headed to infinity and beyond. Perhaps even more tellingly, no effort has been made here to detail the myriad ways that citizens provide data to the marketplace by posting to Facebook, Twitter, Instagram, Foursquare, and other social media sites.

As mentioned, one area of constitutional protections shows tentative promise for providing privacy advocates and citizens with effective tools to combat illegitimate data collection. Recent Supreme Court rulings appear to provide potential tools that could reinvigorate constitutional protections of privacy. These approaches are introduced here and also included in recommendations appearing in chapter 7.

Article III: No Harm No Foul? Perhaps Not So Much

The 2016 Supreme Court decision in *Spokeo v. Robins* sets an important precedent with regard to a broad range of privacy issues, especially on the Internet. The fact that the court accepted the case (in April 2015) and heard arguments (in November 2015) speaks to the importance of the issues at hand. That the case reached the Supreme Court based on a ruling by the Ninth Circuit reversing a district court finding clearly indicates that, regardless of the high court's ruling, the plaintiff's position represents a viable claim to important privacy principles. The specific issue at hand in the case is this: "Whether Congress may confer Article III standing upon a plaintiff who suffers no concrete harm, and who therefore could not otherwise invoke the jurisdiction of a federal court, by authorizing a private right of action based on a bare violation of a federal statute" (*Spokeo, Inc. v. Robins*, "Petition" 2014). The details of the case are as follows:

> Spokeo, Inc. (*Spokeo*) operated a website that provided information about individuals such as contact data, marital status, age, occupation, and certain types of economic information. Thomas Robins sued Spokeo and claimed that the company willfully violated the Fair Credit Reporting Act (FCRA) by publishing false information about him on the website. However, Robins was unable to allege any "actual or imminent harm," so the district court granted Spokeo's motion to dismiss for lack of subject-matter jurisdiction and Robins' lack of standing under Article III of the Constitution. ("*Spokeo, Inc. v. Robins*," *Oyez* n.d.)

Robins appealed the decision, and the Ninth Circuit reversed the district court under the premise that violating the statutory right is sufficient injury to qualify for litigation standing.

The broad question at hand here is the degree to which citizens can claim that their privacy has been invaded by showing that their personal data has been recorded, gathered, and misappropriated regardless of their being unable to show some specific damage from that data collection. Under the status quo, commercial operations that collect data can simply respond, for example, "What's the harm in targeted marketing?" Since users have difficulty indicating the specific harm in the practice, courts are unwilling to hear cases claiming that the data gathering and use is inappropriate.

The Supreme Court's decision could mark a turning point by indicating that the data management practices themselves can be adequately abusive to merit litigation (McMeley et al. 2015). In addition to establishing questions regarding privacy and the use of personal data, the ruling affects the ability of class-action plaintiffs to join together against the data collection practices that are abusive, even when those practices may not have specifically harmed each and every member of the class.

Personal Data and the Value Equation

The data marketplace thrives on an unfair and inaccurate value equation about personal data. Industries using personal data to produce many billions of dollars in annual profits operate under the fiction that each piece of personal data has little or no value and that value only exists after and as a result of aggregating data into sets for analysis and repurposing. However, the reasons for, and results of, gathering and analyzing consumer data give lie to the fiction: Marketers and advertisers thrive on the ability to return targeted appeals directly to the right individuals; intelligence and law enforcement officials seek to identify specific criminals and security risks. The value of aggregated data is made manifest when it is turned upon the individual person, and such events are increasingly common as new media technologies mature.

Commercial entities doing business on the Internet did not transition from the so-called static Web 1.0 to the so-called interactive Web 2.0 accidentally. After the federal government withdrew funding for the Internet, commercial entities that picked up the slack required profit strategies that justified investments. Advertisers, ad sales outfits, and digital service providers alike were unhappy with the lack of effectiveness of banner advertising and with the inability to connect the dots among readers/users, the money spent on advertising over the Internet, and the return on investment (RoI).

In 1994, Netscape employees Lou Montulli and John Giannandrea developed computer code that worked in coordination with online shopping cart functions: The "cookie," a file placed on the visitor's computer, enabled sites to identify return customers and specify aspects of their click behavior (see Turow 2011: chap. 2). Additional tracking technologies soon followed, and the relatively static Web 1.0 was poised for increased economic productivity.

The concept of "Web 2.0" began with a brainstorming session at a conference organized by new media development and training guru Tim O'Reilly. O'Reilly and Web pioneer Dale Dougherty encouraged a group of Web developers and operators to gather and leverage increasing amounts of user data in efforts to encourage users' sense of engagement and participation with websites.

> The race is on to own certain classes of core data: location, identity, calendaring of public events, product identifiers and namespaces. In many cases, where there is significant cost to create the data, there may be an opportunity for an Intel Inside style play, with a single source for the data. In others, the winner will be the company that first reaches critical mass via user aggregation, and turns that aggregated data into a system service. ... A further point must be noted with regard to data, and that is user concerns about privacy and their rights to their own data ... as companies begin to realize that control over data may be their chief source of competitive advantage, we may see heightened attempts at control. ... Users Add Value: The key to competitive advantage in internet applications is the extent to which users add their own data to that which you provide (O'Reilly 2005: 3, sidebar).

Individualized data have enormous value but the marketplace (and our legislatures, our courts, and commercial enterprises) find it convenient (and profitable) to uphold the myths that deny individuals profits from giving up their personal data.

It is impossible to establish accurate financial figures that encapsulate the entire data marketplace. Partial enumerations help us imagine its enormous value. In April 2012, the UK's Centre for Economics and Business Research (Cebr) published *Data Equity: Unlocking the Value of Big Data*, an analysis of the value that increased adoption of big data analytic technologies would bring to the UK economy. The report estimates that "data equity was worth £25.1 billion to UK private and public sector businesses in 2011. Increasing adoption of big data analytics technologies will result in bigger gains, and we expect these to reach £40.7 billion on an annual basis by 2017" (Cebr 2012: 4).

> The main efficiency gain to the UK economy is contributed through improvements in customer intelligence. Consumer spending accounts for over 60 per cent of UK GDP, meaning that enhanced customer intelligence, informed by big data, will have a significant impact on the national economy. Data-driven improvements in targeted customer marketing, the more effective meeting of demand and the analytical evaluation of customer behaviour is forecast to produce £73.8 billion in benefits over the years 2012–17. (Cebr 2012: 5)

While big data will provide economic improvements across numerous UK sectors (manufacturing, government, health care, etc.), the value increases using information provided via the commercial data marketplace in the consumer spending categories alone are enormous.

In the United States, "the market for online advertising—worth $50 billion in 2014—has roughly quadrupled in size since 2005" (PricewaterhouseCoopers 2015). It is estimated that ads placed on mobile apps will increase to $21.2 billion by 2016 and that US Internet advertising will rise to $80 billion, also by 2016 (McChesney 2013: 148). As Michael Rigley notes in his animated short, *Network*, the average person in the US contributes 786 pieces of data a day, resulting in 1.3 million data points in various Web service providers' possession over a 45-month period (the average length of retention). These data contribute to the $34,000 by which the information sector increases in value every second (Rigley 2011).

In short, the notion that the data that drive profits in the marketplace have no value is pure fiction. Although it is difficult to parse the value of a single data point, a growing number of enterprises are responding to the measurement challenge, aiding their clients and customers in leveraging the value of personal information. Although much of the information that these enterprises provide also serves promotional functions for their services, one can glean a sense of the value of non-aggregated personal data from their presentations.

Personal.com was a website enterprise that helped users track the ways that online businesses collected and utilized personal data. Personal.com claimed that people could earn thousands of dollars a year because they "believe companies that earn your business (and those who don't) will be willing to compensate individuals for having the chance to interact with qualified buyers of their particular good or service" (Green 2012). Personal.com no longer offers this service; its business model has changed to data aggregation functions for group work in organizations.

Unfortunately, it isn't easy to recover the potential value from the collection of personal data. Many of the startups claiming to return money to consumers—including Personal.com, Datacoup, and Handshake—use brokerage-like business models that ask users to disclose their online behaviors and accounts. The company returning the profit then cuts out ad sales middlemen by selling that data directly to third parties at negotiated rates (Ehrenburg 2014). This approach might work for consumers who have given up on the concept of privacy as a default right; however, to those who resist giving up their private data in the first place, recovering small amounts of money may seem a pyrrhic victory at best. Still, these and other approaches that measure the value of information make clear that the personal data collected and transacted in the data marketplace have value, whether or not they are aggregated. One can only speculate as to the total value of each person's data, over time. We can, however, be sure that online services like Facebook are not really 'free,' as they may extract hundreds of dollars of value, per year, from each user (Woods 2013).

Recent revelations by Facebook—in the company's 2015 fourth-quarter earnings report—more than suggest that individual users' data have value and that enterprises in the data marketplace can keep track of that value. With 1.59 billion users worldwide, annual revenue of $5.8 billion, and $1 billion in profit, Facebook reports that it makes $13.54 per quarter from the data provided by each US Facebook user (Tsukayama 2016). At first glance, $54.16 in a year may not seem a striking amount. But one needs to keep in mind that data are usually a non-rivalrous commodity. In other words, if Facebook makes a profit from a piece of data, other outfits in the data marketplace can (and do) still make additional profits from the very same data point.

As a mental exercise, let's assume that Facebook is among the best at leveraging the value of a data point. After all, it's one of the biggest data management enterprises in the world. Let's assume that Google, Amazon, and your mobile phone provider are (roughly) equally adept. That puts the annual value of your data to the 'big four' at around $200. For the sake of argument, we'll pretend that other enterprises that leverage your data are not as good at it as are the big four. Let's do some counting and some speculative math.

First, how many social networks (in addition to Facebook) do you use, at least semi-regularly? I don't use social networks, but if I did, I'd probably be on Instagram, Twitter, Snapchat, and Foursquare. Let's say you are on those and on three others I have not named. Now have a look at the bookmarks in your favorite browser and do an honest count of the websites you visit regularly (daily or multiple times a day). Although I may go to more or fewer in a given day, I counted about 21 commercial sites that get a visit from me each day. I run software, as browser extensions, that give me a count of how many (and which) data collection outfits request data each time I visit a site (Privacy Badger and Disconnect). I set them to block cookies and trackers and to deny data requests, but they still give me counts. On average, the commercial sites I visit include eight such requests (this is a non-calculated, conservative, very rough estimate, as some sites make *many* more requests and others make fewer). We will, for the moment, ignore the fact that when these sites collect data from me, they end up passing that data to a raft of other outfits that don't show up in this count (remember the basics of network effects: add a node to the network and the network increases exponentially, not merely additively).

So we add 7 (social media sites) to 21 (commercial sites) and multiply by 8, then by $25 each, coming to a total of $5,600 of value to add to the $200 from the big four. This illustrates a few principles. First, your personal data have a lot of value. In terms of the total revenue generated by using your data in the marketplace, $5,800 annually is most likely just the tip of the iceberg. Second, so-called free online services aren't. We are simply unaware (most of the time) as to how much we are paying into the system in support of the services we use. Third, our guesstimates are merely 'opening bids,' because we have not factored the network effects present within the vast data marketplace. The data points you generate can be (and are) recursively sold. And finally, as suggested by the information that Facebook included in its quarterly financial reports, since this is all based on computers and networks, and since digital leaves tracks and computers are really good at running the numbers, enterprises *can* (or could) account for the value they take from you and/or control (or block) themselves or others from attaining those values. The notion that individual data points—provided by (taken from) your use of digital products, services, and networks—have no individual value is a myth.

· 2 ·

CENTERINGS

> Just because perfect control is not possible does not mean that effective control is not possible. … A fundamental principle of bovinity is operating here and elsewhere. Tiny controls, consistently enforced, are enough to direct very large animals.
> —Lawrence Lessig, *Code, Version 2.0* (p. 88)

The FIPs provide the 'center that holds' for this book. Additionally, a number of conceptual orientations inform the approach represented here. This chapter describes and discusses a number of theories and ideas that are, primarily, directed toward examinations of control aspects in mediated environments, including questions related to privacy and surveillance. Perspectives offered by Robert McChesney, Helen Nissenbaum, Yochai Benkler, Lawrence Lessig, and José van Dijck receive primary attention. From a broader view, questions posed by Neil Postman toward technologies—especially when interrogating new technologies—permeate the discussion throughout. Marshall McLuhan's ideas animate this book.

Theoretical Orientations

Although it is wise to assume that the golden mean should prevail, democracy requires autonomy, autonomy requires privacy, and capitalism requires profit.

The connections among autonomy, privacy, and democracy are fundamental, long-standing, and well established. Individuals whose movements are constrained are imprisoned; individuals whose actions are under constant surveillance are likewise shackled; individuals whose range of action is so minimized as to be ineffective are unable to fully participate in democratic processes. American colonialists revolted against the British largely to gain autonomy. Democratic participation requires reasonable amounts of freedom of choice, movement, and action. Too much autonomy breeds anarchy and so, in the United States, the government plays a role in limiting autonomy to (hopefully) manageable amounts.

Likewise is the case with regard to capitalism and profit. Although one might adhere to the fiction that 'information wants to be free,' capitalism requires profit and new media businesses require capital. We often think of digital products and services as having lower costs than their analog counterparts; only sometimes is that assumption true. Mediated products and services require a wide range of investments and capital outlays. Employees must be paid, electricity and other utility bills must be met, technical infrastructures—including servers, computers, and networks—do not construct or maintain themselves without costs. Shareholders demand returns on investments. Inventors and entrepreneurs thrive on rewards and those benefits are often based on monetary gain.

As noted by McChesney (2013: 23–62), capitalism does not guarantee democracy. Historic developments in the United States find democracy rent with social ills connected to wealth inequality (see Keister 2014 on the recent state of wealth inequality). New media technologies may well exacerbate this trend (McChesney 2015). Critics argue that participants in the new media environment can and should take action to moderate polarizing economic forces. Again, the golden mean is an appropriate starting place. To this end, the current volume examines five aspects that threaten democracy by undermining privacy: government secrecy, commercial invasiveness, consumer carelessness, the new media infrastructure, and reliance on benevolent leadership.

US citizens expect the government to protect them from enemies and security threats. In doing so, the government will use secrecy (intelligence work and spying) as part of its strategy. However, we citizens hope that the government focuses its efforts on our enemies rather than turning a watchful eye on us.

Our participation in a capitalist economy leads us to expect that commercial operations will market and advertise products. We know that increased amounts of information about consumers can help sharpen and target messages in ways that help businesses sell products and services effectively. However, we are not comfortable with commercial operations gathering digital dossiers about each of us and misusing this information in an unregulated data marketplace (that sometimes includes sharing information with government entities).

Regardless of their orientation toward government or commercial actions in these matters, far too many Americans exhibit far too few actions and habits that protect their privacy, preferring instead convenience and entertainment both ready-to-hand.

Following McLuhan, while one could argue, without ever reaching closure, about the right- or wrong-ness of various government and/or commercial data collection, analysis, or marketplace uses of private data, such arguments (about the content in media) often miss the more important theoretical and practical point: The new media infrastructure supports ubiquitous surveillance and data collection.

Finally, US citizens have depended on benevolent leadership as a bulwark against repression and as an excuse for not being more vigilant over their privacy. Many Americans adhere to privacy philosophy that depends on data collection being a neutral or benign force. They think, "Since I'm not doing anything wrong, I have nothing to worry about." Or, "I'm not really concerned about targeted marketing; I don't pay much attention to advertising, and sometimes I get good deals from advertisers." In effect, these attitudes depend on a sociopolitical environment in which leadership is generally benevolent. However, it is not difficult to imagine the emergence of somewhat more sinister or less moderate political leadership. If an intemperate leader (or leadership group) were to emerge, the presence of a new media infrastructure that strongly supports ubiquitous surveillance and data collection might seem decidedly unappealing to average citizens.

In short, our current new media environment threatens our democracy. A number of factors are to blame—including the government, commercial enterprises, and our own behaviors—and a large number of modifications will be required to reset our balance.

The examinations of privacy in this volume assume neither that capitalism is innately corrupt nor that American capitalist democracy is beyond repair. Capitalism is not put forth as a panacea. Rather, the proposals that end this book and the examinations that lead to them seek balance between autonomy and economic viability. Changes must be made within the socioeconomic environment that enable everyday people to regain freedom of choice, movement, and action as they utilize computational, networked communication resources when those products and services meet their wants and needs. Unfortunately, the current state of affairs is neither in balance nor indicative of robust and healthy economic and communication environments.

Interrogating New Media

Following McLuhan, although it is useful to carefully examine the media infrastructure in which one finds oneself, doing so is very difficult in that many of the effects of the dominant medium become environmental and taken for granted. Postman (1999) proposed that rather than acquiesce to, and remain unaware of, the forces that dominant media impose, one should interrogate the circumstances presented (especially) by newly anointed dominant media. He asks:

1. What is the problem to which this technology is a solution? (1999: 42)
2. Whose problem is it? (p. 45)
3. Which people and what institutions might be most seriously harmed by technological solution? (p. 45)
4. What new problems might be created because we have solved this problem? (p. 48)
5. What sort of people and institutions might require special economic and political power because of technological change? (p. 50)
6. What changes in language are being enforced by new technologies, and what is being gained and lost by such changes? (p. 53)

I hope that this volume provides answers to these questions. At the beginning, though, it is useful to at least speculate as to orientations and insights the questions suggest. Such musings draw attention to my general leanings.

At its inception, networked, computational communication was thought by many (perhaps even by some of its inventors) to be a potential solution to problems of information insufficiencies. Whether users were particle physicists wanting to exchange data among and between the main frame machines of like-minded researchers in far-flung places or everyday citizens hoping to exchange recipes or fantasy-sports-team rosters and statistics with other enthusiastic cooks or sports fans, there was a sense that new media (especially the Internet) would solve problems associated with lack of access to information.

Postman was skeptical of this claim, making the counterclaim that the need for information, especially in the United States, had been (for all intents and purposes) solved by the public library system (and Carnegie's contributions to it). Add the information resources later provided by mass media, and Postman wondered aloud whether there really was an information problem for new media to solve. Further, and more specifically, Postman might answer questions 1 and 2—"What is the problem to which this technology is a solution?" and "Whose problem is it?"—by noting that development of a global communication network that enabled, say, particle physicists to exchange data previously isolated in their mainframe computers solved a problem unique to particle physicists and did not, thereby, imply that every human on the planet needed (or would benefit from) networked, computational connectivity. While one can easily suggest other groups in need of the benefits offered by the Internet and facing other so-called problems that connectivity can solve (my focus on particle physicists is hyperbolic and anecdotal for the sake of illustration), it is also apparent that a large number of people use the Internet in ways that do not appear to solve real problems in their lives. In fact, many people use new technologies, including the Internet, in trivial or even harmful ways.

As foreshadowed by questions 3, 4, and 5, Postman would not be surprised that the domination of American culture by new media has inverted the information pyramid. It seems a natural implication of his questions that users now provide the bulk of the informational load carried within new media and the information they provide has great value to corporations and governments, but doesn't result in equal monetary return to users.

Initially, the intended benefits found users gaining information (and control over information) via new media technologies. Unfortunately, we find in the present what early new media innovators and advocates might have seen as unintended consequences: Much of the value in the system is provided by individuals who, largely unwittingly, put information into the data marketplace, thereby providing profit for corporations rather than returning measurable monetary value to users. Although I am not discounting the perceived rewards that users receive from new media (connectivity to others, ease of shopping, knowledge gained, time and/or money saved, etc.), the loss of autonomy and control entailed in the abrogation of privacy brings about numerous new problems and damages real people. Being entertained, staying in mediated contact with others, improving the ability to shop enthusiastically, and being able to look up the answers to an infinite number of (often trivial) questions are, by some measures, advantages that pale in comparison to the damage the loss of privacy does to American democracy and to its now-constantly-surveilled citizens.

Postman encouraged critical thinking about the effects of dominant media. His questions and their answers are kept very close to the surface in this volume.

The Framework of Contextual Integrity

Nissenbaum and others have noted the importance of context in considerations of acceptable or sought-out levels of privacy and surveillance. For Nissenbaum, the framework of contextual integrity examines "finely calibrated systems of social norms, or rules, [that] govern the flow of personal information in distinct social contexts (e.g., education, healthcare, and politics)" (2010: 3). Examining contexts by focusing on technical legal matters (for example, constitutional amendments interpreted to protect privacy), entrenched business practices or industry standards, or assertions about technological imperatives encourages one to construct generalized positions about privacy and surveillance that do not comport with everyday realities. These inflexible views might, for example, propose disallowing secret surveillance. Yet, there are circumstances under which keeping track of certain people (for instance, terrorists or criminals) is a legitimate law enforcement activity. As Nissenbaum points out, what really matters is how the given data collection/ surveillance event comports with social norms or rules that govern the collection, management, and dissemination of information. General and inflexible approaches do not adequately consider norms and rules that are local in both time and geography. Approaches that remove social considerations from judgments about contextual integrity replace users' expectations for informational privacy norms with formalities that lead to surprised, shocked, and annoyed users. For instance, users can feel violated or controlled by formalities (such as unreasonable ToS) or feel that the formalities do not adequately protect them (for example, when the US intelligence community sweeps up the metadata from everyone's phone and text records).

Contextual integrity, explains Nissenbaum, "is achieved when actions and practices comport with informational norms. It is violated when actions and practices defy expectations by disrupting entrenched or normative information flows" (2015: 157). Under the framework of contextual integrity, then, positive regard for entrenched norms becomes the hallmark for judging the propriety surrounding privacy and/or surveillance:

> Although it remains crucial to the understanding of these disputes that we grasp the configurations of interests, values, and principles present in them, our capacity to

explain them is diminished if we attend to these elements alone, blind to the sway of social structures and norms. Tethered to fundamental moral and political principles, enriched by key social elements, the framework of contextual integrity is sufficiently expressive to model people's reactions to troubling technology-based systems and practices as well as to formulate normative guidelines for policy, action, and design. (Nissenbaum 2010: 11)

In order to evaluate the morality of information practices, "one considers the interests of key affected parties—the benefits they enjoy, the costs and risks they suffer. ... A second layer considers general moral and political values. ... Finally, we must consider context-specific values, ends, and purposes" (Nissenbaum 2015: 157–158).

The current volume proposes that the FIPs should be treated as important indicators of the norms that orient examination of these complex questions. There is, of course, a caveat to be noted and an argument to be made against this proposal. Strict adherence to Nissenbaum's characterizations rules out dependence on the FIPs, because they are generally ignored by most parties during common, everyday digital exchanges. As such, it seems that they cannot serve as examples of Nissenbaum's "entrenched norms."

As a result, my claim proposes that they *should* provide needed norms, not that they regularly do so. Critics might argue that few Americans (and especially *not* American businesses or the US intelligence and law enforcement communities) have accepted them and, therefore, they cannot serve as entrenched norms. However, the position taken in this volume is that the various reasons they have not become established are neither fixed nor impenetrable. In fact, the FIPs provide an excellent basis for what should become entrenched norms in the digitally mediated communication environment, thereby helping us make case-by-case assessments of the legitimacy of data management practices within the framework of contextual integrity.

The FIPs are neither formal legal principles nor universals. Rather, they are reasonable and commonsense-based approaches to protecting citizens' privacy from the overreaches of both government and commercial interests. The proof of their validity in this regard is found in their native reasonableness as well as their flexibility across a variety of contextual considerations—the very illustration of a useful norm. The fact that they continue to appear (nearly 40 years after their initial articulation), in modified forms, in virtually every major proposal to protect privacy is evidence of their native reasonableness in light of principles of democracy in the United States. The fact that they have not yet been fully adopted establishes the FIPs as noble goals rather than as markers of failure.

It seems to me to be reasonable (for example, taking the first principle) that if we are to protect citizens' rights, systems for surveillance (data collection, storage, and exchange) should not be secret. A contextual approach recognizes, on the one hand, that it is sometimes reasonable to engage in secret surveillance of a suspected criminal or terrorist; on the other hand, however, it seems relatively unreasonable to secretly observe and record the Web-surfing behavior of all ordinary citizens. The excuses for this practice, often given by industry functionaries—"It's just how the Web works," or, "It's accepted industry practice," or, "How else are we going to authenticate your account?" or, "We do it to improve your experience with our products and services," or, "We warned you in our terms of service"—all seem hollow and diversionary, focused mostly on keeping users in the dark about what's really being done with their private data behind the secret, coded curtain.

While there are situations that would limit the reach and range of the FIPs, their general applicability as indicators of common social norms seems reasonable despite the fact that the principles are neither ensconced in law nor regularly followed by either government or private industry. Sometimes progressive societal change is predicated on what should be the case rather than on settling for what is in place.

Identifying Loci of Control

In *The Wealth of Networks: How Social Production Transforms Markets and Freedom*, Yochai Benkler presented a model for what he termed the "normative characteristics of a communication system"—that is, those aspects that determine "who gets to say what, to whom, and who decides" (2006: 392). He identified a model with three "layers" (physical, content, and logical):

> The physical layer refers to the material things used to connect human beings to each other. These include the computers, phones, handhelds, wires, wireless links, and the like. The content layer is the set of humanly meaningful statements that human beings utter to and with another. It includes both the actual utterances and the mechanisms, to the extent that they are based on human communication rather than on mechanical processing, for filtering, accreditation, and interpretation. The logical layer represents the algorithms, standards, ways of translating human meaning into something that machines can transmit, store or compute, and something that machines process into communications meaningful to human beings. These include standards, protocols, and software—both general enabling platforms like operating

systems, and more specific applications. A mediated human communication must use all three layers, and each layer therefore represents a resource or a pathway that the communication must use or traverse in order to reach its intended destination. (Benkler 2006: 392)

When presenting this model, Benkler focused on the benefits of keeping these layers open and free from proprietary constraints. He was not alone, in the early years of new media, in positing the potentially liberating aspects of distributed, networked, digital communication technologies.

Lawrence Lessig, among others, later adapted Benkler's treatment of layers. Initially, Lessig reordered the three layers by placing the physical layer at the bottom, adding the term "code layer" to what Benkler had called the logical layer, and placing that code layer in the middle of the configuration. Finally, he positioned the content layer—"the actual stuff that gets said or transmitted across these wires" (Lessig 2001: 23)—at the top. In later work, Lessig returned to describe the layers that make up the various parts of the modern Internet (2006). Here, he presented four functional layers residing atop the physical layer in the TCP/IP architecture: the data link, network, transport, and application layers (Lessig 2006: 144). Additional "boxes," representing procedures and content, appear at various nodes throughout and at both the sending and recipient ends of exchanges over the Internet.

In a way that complements Benkler's use of the layer model to indicate the openness available to new media, Lessig demonstrates how open-source implementations are less subject to governmental control than are proprietary solutions. Also, because his approach is cautionary (rather than effusive or optimistic), Lessig shows that controls can be executed within closed networks and argues that the Internet has become more closed than open across its development and maturity.

The current volume illustrates numerous ways that controls are exacted on the various layers described by Benkler and Lessig. In fact, one of the principal claims of this volume is that everyday communication that was previously open and relatively free from control is now closed and subject to the constraints brought about by data gathering and distribution of private information within the data marketplace. Although it is obvious that computationally based communication features code, as Benkler and Lessig noted, it is less obvious that mediating communication via code must always include the loss of privacy and increased surveillance. Early iterations of the Internet featured anonymity and the promise of increased freedom of action. Subsequent iterations have used coded constraints within a variety of the layers described above to close and constrain via loss of privacy and increased surveillance—which is more indicative of the ways North American capitalism works than constitutive of the ways the Internet and new media must operate.

Ideological Powers in Platforms

Numerous authors, including Ken Auletta (2009), David Kirkpatrick (2010), Johnny Ryan (2010), Joseph Turow (2011), and José van Dijck (2014) detail the longitudinal development of new media, including, and especially, aspects related to the Internet. Similarities abound among their descriptions. Most notably, one sees both gradual and sudden movements from open to closed, noncommercial to commercial, and public to proprietary networks and applications as governmental financial (and technical) supports withdrew and private enterprise increased its involvement. Specific applications and platforms, such as Google and Facebook, followed similar life paths as they moved from products and services that initially focused on core functions and values (searching, social networking) to profit-driven enterprises fueled by advertising revenue, driven by high expectations/demands for extensive returns on investment, and reliance on collecting private user information for commercial gain within the data marketplace.

Van Dijck's account focuses on the relationships among everyday users, so-called sociality, industry-based business plans, and control aspects within technological structures.

At its inception, the Web operated primarily as a "noncommercial, public space where they [users] could communicate free of government or market constraints" (van Dijck 2014: 14). Early enthusiasts extolled the virtues of the Web's promises to promote community over commerce, and the early years online featured the relatively peaceful coexistence of commercial and noncommercial platforms.

However, corporate demands for RoI, the enormous profit potentials presented by almost limitless numbers of new users as targets for advertising and marketing, and extensive infrastructure costs required by exponential growth of networks and users of particular applications encouraged Silicon Valley developers to translate the meaning of the terms 'relationships' and 'connectivity' from their social to their commoditized meanings. According to van Dijck, "under the guise of connectedness they produce a precious resource: connectivity. Even though the term 'connectivity' originated in technology, where it denotes computer transmission, in the context of social capital media it quickly assumed the connotation of users accumulating *social* capital, while in fact this term increasingly referred to owners amassing *economic* capital" (2014: 16; italics in original). In effect, Web 2.0 commoditized the interactions that took place on Web 1.0 by converting user actions into economically viable data. The 'trick' of the 'interactive Web 2.0' was its ability to encourage increased use of, and then dependence on, so-called social media as a way to collect and monetize massive amounts of user data—mostly beyond the conscious awareness of the users producing the information, thereby thwarting users' effective control over their own, private, information.

Van Dijck describes the platform infrastructure that accomplishes this trick via five technological dimensions: (meta)data, algorithm, protocol, interface, and default (2014: 30). These five technological dimensions of new media platforms interact with users, content, and the various structures of ownership (business models and the like) to produce profits for commercial entities under the guise of participatory digital communication networks. Van Dijck notes that *data* "can be any type of information in a form suitable for use with the computer," while *metadata* "contain structured information to describe, explain, and locate information resources or otherwise make it easier to retrieve, use, or manage them" (2014: 30). She explains the other dimensions as follows:

> Apart from their ability to collect (meta)data, the computational power of social media platforms lies in their capability to include *algorithms* for processing data … algorithms infiltrate a social (trans)action by means of computational data analysis, upon which the outcome is translated into a commercial-social tactic. Besides deploying algorithms, a platform's coded architecture makes use of *protocols*. Protocols are formal descriptions of digital message formats complemented by rules for regulating those meanings in or between computing systems. … [P]rotocols are technical sets of rules that gain their usability from how they are *programmed* and how they're *governed* or managed by their owners. … Protocols hide behind invisible or visible *interfaces*. Internal interfaces are concealed from the user, who can only see the front end, or the visible interface. An invisible, internal interface links software to hardware and human users to data sources. … Platform owners, evidently, control the internal interface; changes they make to the internal interface do not necessarily show in the icons and features visible to users. … Finally, interfaces are commonly characterized by *defaults*: settings automatically assigned to a software application to channel user behavior in a certain way. Defaults are not just technical but also ideological maneuverings; if changing the default takes effort, users are more likely to conform to the site's decision architecture. (van Dijck 2014: 30–31, 32; italics in original)

These technological features of platforms provide the essential building blocks for new media, enabling us to locate the various ways and places that privacy and surveillance are manipulated and accomplished in digitally mediated environments.

McLuhan's assertion that the medium *is* the message highlights the importance of the surveillance infrastructure being constructed within the new media environment. Postman's assertion that mediums *shape* messages reminds us that some content does not 'play' well within every medium. For example, privacy does not operate efficiently within the new media environment. Nicholas P. Negroponte (1995) pointed out that, in the digital realm, code constitutes both the medium and the message. The ways that we use digital affordances construct and display the new media environment in which both privacy and its absence operate. Chapter 3 examines the secrecy present in the ways that collected private data is leveraged, revealing processes that violate the first FIP.

· 3 ·
FIP 1

No Secret Data Collection

We have no secrets/We tell each other most everything.
—"No Secrets," Carly Simon

There must be no personal-data record-keeping systems whose very existence is secret.
—FIP 1, *Records, Computers, and the Rights of Citizens*

According to Merriam-Webster's Dictionary, the word "secret" does not mean the absolute absence of knowledge about a given subject. The word's definitions focus on purposive subterfuge to hide. Specifically, something that is secret is "kept hidden from others; known to only a few people" ("Secret" n.d.).

Full definition
1.
a: kept from knowledge or view: hidden
b: marked by the habit of discretion: closemouthed
c: working with hidden aims or methods: undercover <a *secret* agent>
d: not acknowledged: *unavowed* <a *secret* bride>
e: conducted in secret <a *secret* trial>
2. remote from human frequentation or notice: secluded
3. revealed only to the initiated: esoteric
4. designed to elude observation or detection <a *secret* panel> ("Secret" n.d.)

Under these definitions, the preponderance of data collection in the digital information environment is done in secret. In fact, merely by reorganizing the definitions, with very few changes, one could write a robust description of the ways in which personal information is collected from and about individuals who use the Internet, mobile phones, credit cards, and other activities that produce digital data as a by-product, as follows:

> Normally, in the U.S., the collection (and marketing) of personal data generated by users when they use the Internet, mobile phones, and credit cards, is kept from their knowledge/view via hidden methods designed to elude observation or detection, in ways that are out of human awareness. Functionaries collecting the information seldom acknowledge their activities; those actions are known only to operatives and revealed only to the initiated. The behaviors of those who are 'in the know' are marked by the habit of discretion.

These characteristics are similar across the data collection and management procedures of both private enterprise and public entities (government, intelligence, and law enforcement). In short, it is very difficult to identify a human activity involving digital media that does not (almost always) violate—in fact, spirit, or both—the first principle of the FIPs.

Private enterprise and public entities, alike, utilize similar data collection and management practices. These procedures contribute to keeping the data collection secret from everyday users in a variety of ways. Details about a number of the most important practices are taken up in this chapter, particularly the following:

(1) The uses of wrap contracts—shrinkwraps, clickwraps, browsewraps, ToS, EULAs, and privacy policies—and the convoluted treatment of these by the courts: Wrap contracts construct and display secrecy in plain view (so-called 'transparency') by working to mimic user consent while providing extensive user data to the marketplace behind the backs of users' conscious awareness or explicit approval.

(2) The roles of data aggregation and so-called anonymity in the marketplace: Certain assumptions led to allowing mass data collection in the age of new media, including the idea that because so much data would be produced, and because so much of it was in one way or another anonymized and treated only via aggregated sets, it was unlikely that particular bits of information would be used against specific individuals (Cairncross 1997). Further, there would be too much data to enable targeting individuals. Therefore, went the argument,

everyday users need not worry about data collection as, at most, it would be used to present generally more efficient advertising, thereby improving the average experience by showing users advertising that is 'relevant' to them. Instead of accruing the benefits of these assumptions of safety in numbers, the realities of the data marketplace are that anonymity has been set aside as massive data sets are mined for targeted marketing, law enforcement, and intelligence gathering purposes, in exceedingly personal and specific ways. Anonymity is dead and aggregation has come to mean more data to analyze and use to target individuals.
(3) Serious difficulties protecting the very *most* private data: personal medical information.
(4) The erosion of legal protections that information and US citizens were afforded in earlier, pre-digital times. In the analog and terrestrial eras (prior to digital new media), privacy was generally protected via tort-related laws that provided punitive protection from invasions of privacy, defamation, and misappropriation of one's image. These torts are based on interpretations of constitutional protections. New media technologies operate in various ways that diminish the applicability of constitutional protections in these areas. Additionally, in the post-9/11 environment, legislatures, courts, law enforcement, and intelligence communities have further denigrated traditional constitutional protections as they apply to new media.

When Is a Contract Just Barely, or Not, a Contract?

Shrinkwraps, clickwraps, browsewraps, ToS, EULAs, and privacy policies are used in order to meet the letter (but not the spirit) of legal obligations surrounding commerce, privacy, and contracts. Although practices vary (a little), the general protocol finds that the collection and management entity (let's call it the service provider) notifies the user of the existence of statements spelling out the arrangements among the service provider, the user, and third parties.

The notifications appear to meet the legal requirements that establish a functional contract, and many courts, across many cases, have ruled in favor of online service providers based on the wrap contracts entailed therein. In concluding her book *Wrap Contracts*, Nancy S. Kim describes why treating online notifications and multi-page legalese as valid contacts is tantamount to judicial malpractice.

> Courts are fond of saying that wrap contracts are the same as traditional paper contracts and that they were applying the same rules of contract law to them. Neither statement is true. ... A refusal to acknowledge the ways in which digital contracts differ from paper ones is simply unrealistic. Wrap contracts take the user further away from actual ascent than contracts of adhesion; it is not merely the issue of nonnegotiability that is at stake. It is the issue of autonomy itself. ... Courts justify wrap contracts by claiming that the non-drafting party manifested consent, but their construction of what constitutes manifestation of consent has wandered too far from the truth. Nonconsent does not become consent simply by claiming it is. Courts, in their zeal to promote commercial transactions in new environments, have enforced unilaterally imposed rules as contracts in order to protect new business models from unfair practices. ... By pretending that wrap contracts are "just like" other contracts—that they were contracts at all—courts established precedent that became difficult to escape. (Kim 2013: 211–212)

Unlike typical face-to-face settings, where clients and service providers exchange, negotiate, and sometimes sign contracts, the typical online environment does not offer opportunities for questions and answers, clarifications, or easy opt-outs. Instead, online, the documentation is (purposefully) overly long and exceedingly technical. Often, the documents are in multiple locations with dubious or difficult-to-find connections among them. Sometimes, documents belonging to multiple entities ("our partners") are involved.

In most instances, it doesn't matter whether or not customers have clicked on the terms and/or read them. In a case decided in July 2015 in the Southern District of New York, the judge—following a long line of precedent—upheld the enforceability of a website clickwrap agreement even though the defendant argued "that he was not even constructively aware of the terms of the agreement because it was only accessible via hyperlink" (Neuburger, "Clickwrap" 2015).

One additional, telling example is sufficient here: Please search Google with the phrase "Google terms of service." Although results will vary (our search results are tailored by Google, based on our past browsing histories), your results may be similar to mine. The list of headings in my search results includes:

> Google Terms of Service – Privacy & Terms – Google
> Terms of Service – Google Analytics – Google
> Google Maps/Google Earth APIs Terms of Service

Terms of Service – Buyer – Google [for using Google Wallet]
Terms of Service – Seller – Google [for using Google Wallet]
Privacy Notice – Google Wallet
Google Cloud Platform Terms of Service
Google Drive Terms of Service – Drive Help
Google APIs Terms of Service – Google Developers
Google Play Legal Information
Google Play Terms of Service
Google Books Terms of Service
Terms of Service – Translate API – Google Cloud Platform
BUSINESS PHOTOS TERMS OF SERVICE – Google Maps
Terms of Service – Google Fiber
Additional Terms: Google Project Hosting.
GOOGLE PLAY GIFT CARD® TERMS OF SERVICE – Google Play
Terms of Service – YouTube
3D Warehouse – Terms of Service
Google Books Terms of Service

It is, of course, clear that Google is a big outfit providing a wide range of services and products across numerous business categories. One expects their wrap contracts to be numerous and complex. Clicking to follow a link to the ToS inevitably sends one to a page containing links to other policies and sites. The company's base/master policy reads as follows:

> By using our Services, you are agreeing to these terms. Please read them carefully. Our Services are very diverse, so sometimes additional terms or product requirements (including age requirements) may apply. Additional terms will be available with the relevant Services, and those additional terms become part of your agreement with us if you use those Services. ("Terms of Service" n.d.)

And so, typical users find themselves pretty quickly enmeshed in a broad range of agreements—agreements that they usually have not read and may not understand if they do (but that the courts treat as binding contracts).

A (not so) simple example illustrates: Many of Google's ToS pages link to its base/master privacy policy. The policy (not counting hyperlinks and pop-up informational bubbles) runs approximately 2,800 words in 68 lines. Much of the information is annotated with links or pop-ups that provide details or definitions. In the section titled "Information we collect," Google refers to two broad categories: information you give us, and information we get from your use of our services. Google notes six types of information that are collected: device information, log information, location information, unique application numbers, local storage, and cookies and similar technologies. While it is possible that the average user understands many of the terms used, and while it is plausible that Google's definitions and pop-ups help folks who do not fully understand, it is clear—even to the expert—that the descriptions provided are detailed and very technical in nature. One particularly important section—"Cookies and Similar Technologies"—serves as an illustration:

> We *and our partners* use various technologies to collect and store information when you visit a Google service, and this may include using *cookies or similar technologies* to identify your browser or device. We also use these technologies to collect and store information when you interact with services we offer to our partners, such as *advertising services* or Google features that may appear on other sites. Our Google Analytics product helps businesses and site owners analyze the traffic to their websites and apps. When used in conjunction with our advertising services, such as those using the DoubleClick cookie, Google Analytics information is *linked, by the Google Analytics customer or by Google, using Google technology, with information about visits to multiple sites*. Information we collect when you are signed in to Google, in addition to information we obtain about you from partners, may be associated with your Google Account. When information is associated with your Google Account, we treat it as personal information. For more information about how you can access, manage or delete information that is associated with your Google Account, visit the *Transparency and choice* section of this policy. ("Terms of Service" n.d. [italics indicate the presence of a hyperlink])

In microcosm, and in effect, this policy serves as an apocryphal instance of the way that the Internet works. Google and its partners—very few people know how many partners it has or who those partners are; identifying them would lead one to additional wrap contracts issued by those parties—use cookies and other tracking technologies to collect, store, analyze, and exchange information about users and their electronic and buying behaviors. Although the words are plain enough, and plenty of definitions and illustrations exist to help consumers understand the nature, breadth, and force of the wrap contracts detailed in the policies, few who read them come away with a clear understanding of what they have agreed to. Most click an "Agree" button or just move along to the content they want, resigned to their acceptance of the terms.

Lawyers specializing in intellectual property and/or contract law go to great pains to write documents that protect the service provider (and its business) by detailing and constraining every imaginable aspect of the provider-client relationship. The consumer is left with two options: to accept the terms (perhaps unread and/or without understanding them) or to stop using the service. Most users never read the terms; they simply click the box that indicates acceptance of the service provider's specifications or continue using the online service (thereby accepting the terms without specifically clicking in a box). Continued use of the services or acceptance of the terms is treated by courts as documentation of binding contracts.

The effect of the documentation is obfuscation of the service provider's data collection and management procedures. The more information is provided, the more obfuscation successfully hides the real functions. In effect, ToS, EULAs, and privacy policies hide the real data collection and management procedures in plain sight. Two additional examples highlight these trends. First, in March of 2015, *ars technica* reported on a third-party extension called Webpage Screenshot designed for Google's Chrome browser:

> It boasted more than 1.2 million downloads ... [and] collected users' browsing habits behind the scenes. The snooping was made harder to detect because Webpage Screenshot didn't start collecting the data until a week after the extension was installed. In fairness to the company that produced Webpage Screenshot, the extension's terms of service disclosed that it collected a wealth of potentially sensitive user data. Data that was fair game included IP addresses, operating systems, browser information, URLs visited, data from URLs loaded and pages viewed, search queries entered, social connections, profile properties, contact details, and usage data, along with other behavioral, software and hardware information and unique mobile device identifiers. (Goodin, "Chrome Extension" 2015)

A user of the Google extension read, and raised the alarm over, the click-through notification indicating that Webpage Screenshot came with wrap contracts. It is highly unlikely that many users read them. And of course, if they had, and had disagreed, their only choice would be to not use the extension, an outcome that at least 1.2 million folks found unsatisfactory.

Second, in August of 2015, *Wired.com* published "You Can't Do Squat about Spotify's Eerie New Privacy Policy":

> Spotify released a new privacy policy that is now in effect, and it turns out that the company wants to learn a lot more about you and there's not much you can do about it.
>
> **Spotify Wants to Go Through Your Phone**
> "With your permission, we may collect information stored on your mobile device, such as contacts, photos, or media files."—Spotify
>
> **Spotify Wants to Know Where You're Going**
> "Depending on the type of device that you use to interact with the Service and your settings, we may also collect information about your location based on, for example,

your phone's GPS location or other forms of locating mobile devices (e.g., Bluetooth). We may also collect sensor data (e.g., data about the speed of your movements, such as whether you are running, walking, or in transit)."—Spotify

Spotify Wants to Be Your Facebook Friend
"You may integrate your Spotify account with Third Party Applications. If you do, we may receive similar information related to your interactions with the Service on the Third Party Application …"—Spotify

So, What Can You Do About This?
Sadly, not a whole lot.
"If you don't agree with the terms of this Privacy Policy, then please don't use the Service."—Spotify. (Gottsegen 2015)

The next day, Spotify's CEO clarified the terms, noting that customers would be asked if they wanted to opt out of a given category of data collection (Barrett 2015). In effect, however, the complexities of the terms and the data collection procedures combine with users' preferences for using Spotify and contribute to users' inability and unwillingness to risk causing Spotify to not work properly (or at all) by opting out of standard (and out-of-view) procedures.

In some situations, software developers and providers are clear about their intentions to collect data behind the backs of users. In other cases, the data marketplace environment is so complex that there are differences between what software developers know about secret data collection and what the corporations that provide software know.

In the first case, for example, Microsoft has made it clear that the Windows 10 operating system collects more information than users suspect (or are used to) and that even though it takes place behind the backs of users, Microsoft intends to continue the practices.

> In the run-up to the launch of Windows 10 earlier this year, users noticed that Microsoft's operating system would be collecting more data on them by default than it had in the past—including information about their location and what they're typing—and sending it off to Microsoft. Understandably, some folks were concerned about the privacy implications of such a move. … Those concerns weren't helped by Microsoft, which was slow to clarify exactly what it takes from users, and to explain how to disable much of that collection. … Windows 10 also currently requires that all users hand over some information about how their devices are being used—what Microsoft calls basic telemetry. That information relates to things like when and how Windows 10 crashes. … [T]he company needs that information to improve the experience of using its operating system. For example, "in the case of knowing that our system that

we've created is crashing, or is having serious performance problems, we view that as so helpful to the ecosystem, and [therefore] not an issue of personal privacy, that today, we collect that data so that we make that experience better for everyone," he said. In Belfiore's view, Microsoft's current setup properly addresses users' privacy concerns, since it lets them opt out of the collection of personal information. And as for the tracking features that users can't opt out of, he said the company doesn't consider them to be a privacy issue. "In the cases where we've not provided options, we feel that those things have to do with the health of the system, and are not personal information or are not related to privacy," he said. (Frank 2015)

In other cases, users' personal information has been collected and put at risk by providers (the right hand) who are not fully aware or in control of what developers (the left hand) are doing as was the case with Apple's struggles over rogue apps in their store.

Data Anonymity via Aggregation as Oxymoron

Increasingly since 2007, knowledgeable researchers and critics of the data marketplace have been clear that it is inaccurate and disingenuous to lead users to believe that data aggregation and anonymity protect them from personalized data analysis and results implementations. Many product/service providers explicitly tell consumers that the personally identifiable data they provide is being used by the company (Facebook, Twitter, Instagram are three good examples), and fewer providers than ever make assurances that anonymized and aggregated information somehow protect users from personalization. Only the most naïve user fails to suspect that targeted marketing must rely on at least some level of personalization.

However, many firms that rely on personal data still encourage consumers to consider, and to believe, the mythic notion that users are not personally identified in anonymized and aggregated information. Users who spend time browsing parts of wrap contracts will encounter nuanced encouragement to accept and adopt antiquated assumptions about the power of anonymization and aggregation. The following examples were taken from the privacy policies of Instagram, Blizzard Entertainment (developer of *World of Warcraft*), and the *New York Times*, respectively:

> We may remove parts of data that can identify you and share anonymized data with other parties. We may also combine your information with other information in a way that it is no longer associated with you and share that aggregated information. ("Privacy Policy," Instagram n.d.)

> We may, however, share non-personally identifiable, aggregated, and/or public information with third parties. ("Blizzard Entertainment," 2015)

> We may transmit non-personally identifiable website usage information to third parties in order to show you advertising for The New York Times when you visit other sites. ("Privacy Policy," *New York Times* 2015)

Reporting in 2007, based on research using the Netflix prize dataset, Arvind Narayanan and Vitaly Shmatikov demonstrated "that an adversary who knows only a little bit about an individual subscriber can easily identify this subscriber's record in the dataset. ... [W]e successfully identified the Netflix records of known users, uncovering their apparent political preferences and other potentially sensitive information." And in 2013,

> Researchers ... analyzed data on 1.5 million cellphone users in a small European country over a span of 15 months and found that just four points of reference ... was enough to uniquely identify 95 percent of them. ... [T]o extract the complete location information for a single person from an "anonymized" data set of more than a million people, all you would need to do is place him or her within a couple of hundred yards of a cellphone transmitter, sometime over the course of an hour, four times in one year. (Hardesty 2013)

One of the roadblocks to fully establishing that anonymized and aggregated information compromises personal privacy is that court decisions still rely on the assumption that anonymization and aggregation erase pointers to individuals. For example, "a federal judge in New York ordered YouTube to provide Viacom with the IP addresses of users, as part of Viacom's copyright infringement lawsuit against the video-sharing site. The judge wrote at the time that IP addresses alone can't identify users." Additionally, "a federal judge in Kentucky ... ruled that a nursing student at the University of Louisville didn't reveal personally identifiable information when she posted information on MySpace about a patient who had just given birth to a baby girl" (Davis 2009).

In May 2016, the US Supreme Court ruled, in *Spokeo v. Robins*, that plaintiffs could litigate for damages from inaccurate Web postings even in cases when they could not prove "injuries in fact." Data collected surreptitiously can lead to inaccuracies and harms to plaintiffs (*Spokeo v. Robins*).

Solove (2004, 2011) and others (e.g., Aftergood 2015) have long—and persuasively—argued that secrecy presents significant challenges to FIPs. In terms of the private data marketplace, many of the key practices endemic to the Internet-based networked commercial environment are cloaked in technological mystery, largely via ubiquity. Citizen-consumers put up with (or ignore) intrusive data collection, for the most part, because they value the so-called 'free' services that appear to improve their quality of life via convenience (Schneier 2015: 201).

But as Solove has repeatedly stressed, perhaps the *most* significant aspect of secrecy and the data marketplace is the degree to which data streams generated by private enterprise overlap with those generated by governmental agencies, without the conscious awareness of citizens.

It is entirely one thing to acquiesce to surveillance in the data marketplace for the conveniences provided by the outcomes of advertising, marketing, and consumerism.[1] Unknowingly providing that data to law enforcement and governmental agencies is quite another matter, and secrecy lends a chilling dimension to the situation (for example, see Kravets, "Surveillance Society" 2015). In effect, US citizens expect the government to collect information about them but they do not expect private enterprise to do that work for the government. Government-based surveillance and data collection, in secret, presents problems of its own.

Further, citizens often benefit from governmental transparency. When the government releases data that can contribute to the public good, doing so is useful and is in accordance with the FIPs. For example, the US Food and Drug Administration (FDA) provides the "Open Government: FDA Data Sets" collection as "a new initiative designed to make it easier for web developers, researchers, and the public to access large, important public health datasets collected by the agency" ("FDA Launches" 2014). However, once released to the public, those databases are also available for use by commercial entities for advertising and marketing purposes as well as other data analytic procedures (Farivar, "Private Firms" 2014). Data provided to the government by citizens for one reason is then repurposed for other reasons out of the citizens' control and in contravention to the FIPs.

For example, in *The Digital Person*, Solove (2004) details the increasing ineffectiveness of constitutional protections against intrusive government surveillance and data collection. Accounts of the Pentagon's and the National Security Agency's spending and building programs appeared in press reports and books (Bamford 2009, 2012).

Even the Most Private Datum Isn't

Information about one's health and health care, sometimes referred to as 'sensitive data,' is thought by many to be the most private of all personal information. HEW's recommendations for sensitive data, the FIPs, were most strongly adopted in the healthcare environment. As noted by Dr. Deborah Peel, "the public believes that the Health Insurance Portability and Accountability Act (HIPAA) protects the privacy of sensitive health data, but the reality is current health information technology systems prevent patient consent and control over personal health data" (2015: 174). Peel continues:

> Personal health data in the United States is bought, sold, and traded by nearly a million health data brokers *millions of times a day, without the knowledge or consent of the individual.* ... Today, healthcare institutions, government, technology vendors, and health data holders treat patient data like a proprietary asset, as if individuals no longer have fundamental legal or ethical rights to control the use of personal health information. ... "[D]ata brokers" make tens to hundreds of billions of dollars in revenue annually. ... The states and federal government sell or disclose personal health data, too. HealthData.gov has released over 1000 data sets of health information for public use, even though the intent of the "open data" movement was to open up data about government, not individuals. All fifty states sell newborn blood spots and thirty-three states sell inpatient and outpatient data. Further, current health technology systems also make it impossible for individuals to know about the millions of daily human and electronic uses of personal health information in health technology systems. ... The "accounting for disclosures" of all patient data from electronic health records, as required by the ... 2009 American Recovery and Investment Act, has yet to be implemented in regulations. Individuals have no access to "chain of custody" so they can track all uses of their most sensitive personal data. (Peel 2015: 175–176)

The scope of data breaches in the medical records environment is stunning. Between September 2009 and August 2012 alone, information from the medical records of about 21 million patients were exposed via "477 breaches affecting 500 people or more each," as "reported to the Office for Civil Rights (OCR) under the US Department of Health and Human Services. In total, the health records of 20,970,222 people [were] compromised" in this three-year time frame (Mearian 2012). A later analysis of government records, published in the *Journal of the American Medical Association* (*JAMA*), found that close to a thousand large data breaches affected 29 million medical records between 2010 and 2013 (Brown 2014).

Secret use of medical information (which is more common than most people imagine) and an increasing ability to de-anonymize data complicates the situation and compromises permissions that had assumed anonymity. Consider the fact that "[t]issues from millions of Americans are used in research" without their providers' knowledge:

These "clinical biospecimens" are leftovers from blood tests, biopsies and surgeries. If your identity is removed, scientists don't have to ask your permission to use them. ... The United States government recently proposed sweeping revisions to the Federal Policy for Protection of Human Subjects, or the Common Rule, which governs research on humans, tissues and genetic material. ... The most controversial change would require scientists to get consent for research on all biospecimens, even anonymous ones. The Common Rule doesn't require consent for "non-identifiable" samples, but scientists have proven it's possible to "re-identify" anonymous samples using DNA and publicly available information. (Skloot 2015)

The federal government continues to examine, and reconsider, rules about the practice of using tissues in research, but in the meantime, a wide range of uses for personal medical data rely on patients' informed consent as an authorizing mechanism. Perhaps no type of wrap contract is as complicated and problematic as medical releases.

As a local illustration, I recently filled out a privacy form at the office of my family doctor. My primary healthcare provider is affiliated with a large healthcare network that includes numerous physicians and hospitals in multiple states. The form provides a mundane yet graphic illustration of the crisis in the private medical information marketplace. In many ways, the form comports with the FIPs. In fact, policy forms of this nature, in the healthcare environment, are supposed to meet both the letter of the law and the spirit of the FIPs.

The form, "Patient Authorization for Release of Information," includes seven major sections: (1) identifying information; (2) contents to be disclosed; (3) term of the permission; (4) purposes for disclosure; (5) methods of disclosure; (6) signature, date, and witness representations; (7) and an informational section that notifies patients about various procedural aspects on the part of the healthcare network and the patient.

In the informational section, the document notifies the patient of six important aspects, some of which accomplish the goals of the FIPs: (1) that information provided to third parties is no longer covered by the authorization; (2) that the medical providing group may be receiving remuneration from a third party in connection with the use of the health information provided; (3) that the authorization can be revoked by way of written notice; (4) that the patient may refuse to sign or may revoke the authorization at any time; (5) that the patient may inspect their health data; and (6) that the patient has read and understood the terms of the agreement.

The third, fourth, fifth, and sixth notifications at least partially provide patients with tools and protocols for examining their health records and potentially making changes in those records. All of the notifications serve the general purpose of letting patients know how the network manages their health data.

However, items one and two serve as a kind of ToS that undermines the FIPs. In number one, patients are reminded that any health data subsequently provided to third parties are not covered by the authorization forms that the patient is signing. None of the specifications or constraints that the patient attaches via the form, or via subsequent written notifications, apply to any of the data once they have been provided to third parties. Further, number two notes that the network may receive remuneration from third parties for making the data available.

The specific wording of the form does not ask the patient whether or not they authorize the sale of their data; rather, it merely notifies them that the network may receive remuneration: "I understand that Methodist [the healthcare provider] may, directly or indirectly, receive remuneration from a third party in connection with the use or disclosure of my health information." Notice how this wording is a form of secrecy, a hiding in plain sight that discourages patients' knowing participation in decision making about their medical data. If the form read "Do you give us permission to sell your medical data to insurance companies and pharmaceutical organizations?" I suspect that most patients would check the "no" box. Instead, the notification, that has just informed patients that information provided to third parties is not protected by the terms of the authorization form that they are signing, appears more as a fait accompli: "We are letting you know that we are receiving money for your health data."[2]

One additional nuance of ToS construction also hides data abuse in plain sight: The offers that specify patients' abilities to revoke authorization—sections three and four—do not enable patients to indicate which portions of the authorization they want to control or revoke. In the end, the form is confusing, contains many boxes to check and variables to circle, asks for multiple decisions, and provides many notifications to evaluate—all while one is presumably sick or injured and waiting for an appointment with a physician—tending to cause most users to simply sign and date the form without a second thought. Again, this form is an example of data management hidden in plain sight and hardly fulfills the FIPs (or does so only technically). The form is also an example of bad construction and execution of a medical ToS.

Recent history is replete with instances of hacking that have compromised medical data (for example, see Brown 2014). Likewise, the ways that medical information is entered into databases and shared across enterprises involve complexities that compromise the ability of patients and consumers to interact within medical communication networks with confidence that their privacy will be protected (for example, see Dwoskin 2014).

Further, a large number of consumers now collect and compile personal medical data and exchange it with app developers and interactive systems in an ever-increasing mobile environment. Charles Ornstein of ProPublica, writing in the *Washington Post*, notes:

> The Health Insurance Portability and Accountability Act, a landmark 1996 patient-privacy law, only covers patient information kept by health providers, insurers and data clearinghouses, as well as their business partners. At-home paternity tests fall outside the law's purview. For that matter, so do wearables like Fitbit that measure steps and sleep, gene testing companies like 23 and Me, and online repositories where individuals can store their health records. In several instances, the privacy of people using these newer services has been jeopardized, causing embarrassment or legal repercussions. (Ornstein, "Federal Privacy" 2015)

FTC Commissioner Julie Brill adds that "[c]onsumer-generated health information is proliferating," but many users don't realize that much of it is stored "outside of the HIPAA silo" (quoted in Ornstein, "Federal Privacy" 2015).

One has to wonder: If we cannot protect even our most intimate healthcare data, and we don't know about or understand the totality of the breaches (potential or intentional), what can we do to protect any, perhaps less sensitive but still important, personal/private information? And if we can't control private information in the commercial data marketplace, what chance do we have in relation to governmental (law enforcement and intelligence community) uses and abuses?

Knowing about the Unknowable

Data provided by Edward Snowden confirmed and extended previously held suspicions about governmental abuses of US citizens' personal and private information, not to mention abuses involving overseas allies and foes alike (Gidda 2013; Greenslade 2013; Savage, "CIA Is Said" 2013). Prior to Snowden's revelations, numerous reports had provided information about implementations of government surveillance programs, particularly since the 9/11 attacks in 2001. While the Snowden materials were wide ranging and raised public awareness by providing extensive and specific evidence, privacy advocates and reporters on the surveillance beat had long warned of increasing buildups of government data collection programs (Bamford 2009, 2012; Whittaker 2012).

Opinions as to whether Snowden is a patriot or a traitor vary widely and are charged with partisan political passion. But even his most strident critics are now apt to admit that his revelations helped to bring important questions forward for public debate (Gallagher, "Director" 2015). Opinions also vary as to whether secretly collecting massive amounts of electronic data generated by American citizens—for the purpose of data analysis to ferret out potential terrorist threats—was, is, or will be legal or in our best interests (Landers 2013).

Regardless of what we've learned about government surveillance, it is clear that much of the surveillance and data collection initiated by law enforcement and governmental agencies takes place within a secrecy paradigm. To some extent, that is to be expected; spying on our enemies will, and perhaps should, happen mostly in secret. Americans are aware, and in many cases supportive, of governmental efforts to protect the homeland from terrorists and lawlessness (Madden 2014). Law enforcement pursuit of criminal activity and homeland security activities directly relevant to terrorism and/or cyber terrorism will also take place out of public view.

Hopefully, these valid governmental functions will abide by the legal constraints and program restrictions that have been put in place legislatively. Obviously, debates arise as to whether our security and surveillance infrastructures have followed the letter and spirit of such constraints. Some of the concerns about secrecy in data gathering and analysis relate directly to deliberate abrogation of these constraints by program participants (Brill 2013; Mullin 2013).

Questions about the first principle of the FIPs, secrecy, as they relate to governmental surveillance and data collection, do not respond directly to a simple inquiry as to whether a given program is secret. The assumption is that many of the programs are and will be secret. The most significant questions about such activities relate to whether secrecy is used as a tool in order to evade legal constraints and reasonable efforts to protect the rights of citizens.

For instance, according to Kevin Bankston, the director of New America's Open Technology Institute, there is a "[n]ew rule: if the NSA claims that a particular surveillance program has ended, or that a particular type of surveillance has halted 'under this program,' assume that it is still going on in another program" (quoted in Peterson, "Why It's So Hard" 2015). Andrea Peterson, technology policy reporter for the *Washington Post*, writes that "big disclosures such as Snowden's come along rarely. And now we're seeing that reporting on these programs is like a sort of like playing Whack-A-Mole: Even if one program appears to have ended, others spring up in their place—and the general public often doesn't learn about them until years after they've taken effect" (Peterson, "Why It's So Hard" 2015).

The use of secrecy as a tool against American citizens violates both the spirit and the letter of the first FIP.

Quick Reminder: The Constitution Is Supposed to Protect Our Privacy

The US Constitution provides a variety of approaches and tools for protecting privacy that have stood as the bulwark of privacy protection in this country. However, in the post-9/11 environment these protections have been undermined by executive, legislative, and (especially) judicial actions combined with the "normal functioning" of the commercial data marketplace (Solove 2004). Among other things, the First Amendment

> safeguards the right of people to associate with one another. Freedom of association restricts the government's ability to demand organizations to disclose the names and addresses of their members or to compel people to list the organizations to which they belong. ... [P]rivacy of associations is becoming more difficult in a world where online postings are archived, where a list of the people a person contacts can easily be generated from telephone and email records, and where records reveal where a person travels, what websites she visits, and so on. The Supreme Court has repeatedly held that the First Amendment protects anonymous speech, and it can restrict the government from requiring the disclosure of information that reveals a speaker's identity. However, the First Amendment only applies when the government plays a role in the compulsion of the information, and most of the gathering of personal information by companies isn't done under the pressure of any law. (Solove 2004: 62–63)

The Fourth Amendment restricts the government from conducting unreasonable searches and seizures, and the Fifth Amendment protects individuals against being forced to incriminate themselves. As with other constitutional protections, these statutes do not affect the way that private enterprise (the data marketplace) collects, analyzes, and redistributes information. In very many ways, government actions and court decisions have strongly undermined most of the protections that would be afforded by these constitutional amendments when applied to electronic surveillance and data collection in the new media environment.

Solove provides an example in the 1995 case of *Doe v. Southeastern Pennsylvania Transportation Authority (SEPTA)* in which "the constitutional right to information privacy failed to comprehend the privacy problem of databases" (2004: 66):

> The plaintiff Doe was HIV positive and told two doctors ... at his work about his condition but nobody else. He strove to keep it a secret. His employer, SEPTA ... maintained a prescription drug program with Rite-Aide as the drug supplier. SEPTA monitored the cost of its program. Doe was taking a drug used exclusively in the treatment of HIV. ... Unfortunately, even though SEPTA never asked for the names, Rite-Aide mistakenly supplied the names corresponding to prescriptions when it sent SEPTA the reports. Doe began to fear that other people at work had found out ... [and] that people were treating him differently. Regardless of whether he was imagining how his coworkers were treating him, he was indeed suffering a real, palpable fear. ... The privacy problem wasn't merely the fact that [someone had] divulged his secret or that Doe himself had lost control over his information, but rather that the information appeared to be entirely out of anyone's control. (Solove 2004: 67)

Traditional protections provided by the Fourth and Fifth Amendments to the Constitution do not protect citizens from many of the abuses that result from data collection and information distribution. For instance, in *Doe v. SEPTA*, the court ruled that the defendants did not violate the plaintiff's right of privacy, because

> a self-insured employer's need for access to employee prescription records under its health insurance plan, when the information disclosed is only for the purpose of monitoring the plans by those with a need to know, outweighs an employee's interest in keeping his prescription drug purchases confidential. Such minimal intrusion, although an impingement on privacy, is insufficient to constitute a constitutional violation. ("*John Doe*" 1995: 19)

The fact that interpretations of traditional protections have changed in the new media environment is only the tip of the iceberg in discussions about the weakening of constitutional protections from secret surveillance and data collection. Enormous government intelligence programs, and various ways that our laws and courts fail to protect citizens, are more dramatic and worrisome. Likewise, the governmental appropriation of private data gathering exacerbates the situation. A few examples illustrate.

As far back as 2002, *New York Times* columnist William Safire published information about the intelligence community's Total Information Awareness program. Safire's writing, and later investigations, led to congressional action, in 2003, to dismantle the program that would have created massive digital databases and dossiers on American citizens by drawing together all available big data resources (Safire 2002; Scheer 2015). And yet, in 2005, when intelligence employee Thomas Drake

> allegedly gave information to the *Baltimore Sun* showing that the publicly discussed program known as Trailblazer [one of the alternative programs set up by the NSA after it had lost Total Information Awareness] was millions of dollars over budget ..., federal prosecutors charged Drake with 10 felony counts. ... He faced up to 35 years in prison—despite the fact that all the information Drake was alleged to have leaked was not only unclassified and already in the public domain but in fact had been placed there by the NSA and Pentagon officials themselves. ... [T]he investigation went on for four years, after which Drake received no jail time or fine. (Bamford 2013: 94–95)

In effect, while the government and intelligence communities were harassing a whistleblower, the protections ostensibly afforded by the Constitution did little to stop the federal government from, essentially, ruining Drake's life. The judge who finally dismissed the charges "[e]xcoriated the prosecutor and the NSA officials for dragging their feet. 'I find that unconscionable. ... That's four years of hell that a citizen goes through. It was not proper. It doesn't pass the smell test'" (Bamford 2013: 95).

While the courts have, laboriously, gone back and forth over the past decade debating the legality of massive governmental data gathering, both the courts and the legislative bodies have repeatedly decided that mass data collection should continue, regardless of whether it violates constitutional protections (against self-incrimination, for example).

No attempt is made in the list that follows to document the current state of affairs with regard to the (il)legality of each practice. By the time you read this book, these matters will have shifted numerous times; there's no such thing as a final score on this scorecard. Instead, the following items are included to illustrate that expectations of constitutional protections of privacy have lessened dramatically during the new media and post-9/11 eras. First, one can note the deluge of government requests for data or data actually collected by governmental agencies.

> "Verizon Says It Received Over 321,000 Legal Orders for User Data in 2013." January 22, 2014.
> Joining the parade of technology companies that are releasing "transparency reports" as a window into government legal pressure, Verizon announced on Wednesday that

it received over 321,000 total orders from various American law enforcement agencies in 2013. (Farivar, "Verizon Says" 2014)

"How the NSA Collects Millions of Phone Texts a Day." November 9, 2014.
The NSA collects 194 million messages a day—not just SMS messages, but system-generated messages as well: geolocation data, synchronizing address book data (vCards), missed call messages, call roaming data, and other data as well. (Gallagher, "How the NSA" 2014)

"When NSA and FBI Call for Surveillance Takeout, These Companies Deliver." September 5, 2014.
Smaller ISPs who aren't frequently hit with warrants can't afford to keep the infrastructure or manpower on-hand to respond to requests—so they sign up with a "trusted third party" capable of doing the work as an insurance policy against such requests. (Gallagher, "When NSA" 2014)

Second, take note of the remarkable back-and-forth both within the judiciary and between the judicial and legislative branches concerning the legality of various data collection programs.

"US Privacy Watchdog: NSA Phone Records Program Is Illegal." January 23, 2014.
The U.S. Privacy and Civil Liberties Oversight Board said in a report ... [t]he NSA lacks the legal authority to collect millions of U.S. telephone records under the Patriot Act. ... The program "lacks a viable legal foundation, implicates constitutional concerns under the First and Fourth Amendments, raises serious threats to privacy and civil liberties as a policy matter, and has shown only limited value" the report said. (Gross, "US Privacy Watchdog" 2014)

"House Okays Cyberthreat Sharing Bill Despite Privacy Concerns." April 22, 2015.
The U.S. House of Representatives has voted to approve legislation that would encourage companies to share cyberattack information with each other and with the government, despite concerns that it would put new consumer information in the hands of surveillance agencies. (Gross, "House Okays" 2015)

"NSA Collection of Bulk Call Data Is Ruled Illegal." May 7, 2015.
A federal appeals court in New York ruled on Thursday that the once-secret National Security Agency program that is systematically collecting Americans' phone records in bulk is illegal. (Savage and Weisman 2015)

"Let the Snooping Resume: Senate Revives Patriot Act Surveillance Measure." June 14, 2015.
The Senate on Tuesday revived three surveillance provisions of the Patriot Act that had expired early Monday because of Senate discord. ... One Patriot Act provision renewed under the bill was a variation of the phone-records spy program that

National Security Agency whistleblower Edward Snowden disclosed in 2013. ... The "business records" section enabled the NSA's bulk telephone metadata program. It grants the government powers to seize most any record, even banking and phone records, by getting a warrant from the Foreign Intelligence Surveillance Act (FISA) Court. ... The second provision revived Tuesday concerns roving wiretaps. Spies may tap a terror suspect's communications without getting a renewed FISA Court warrant, even as a suspect jumps from one device to the next. ... The third spy tool renewed is called "lone wolf" in spy jargon. It allows for roving wiretaps. However, the target of wiretaps does not have to be linked to a foreign power or terrorism (Kravets, "Let the Snooping" 2015).

"Read the Ruling: Appeals Court Reverses on NSA Phone Surveillance." August 28, 2015.
A three-judge panel of the U.S. Court of Appeals for the D.C. Circuit reversed the December 2013 ruling by U.S. District Judge Richard Leon, who wrote that the NSA's "almost Orwellian" bulk metadata collection after the Sept. 11 terrorist attacks "almost certainly" violates constitutional privacy protections (Gershman, "Read the Ruling" 2015).

"DC Appeals Court Lifts Injunction against NSA's Bulk-Collection Program." August 28, 2015.
The National Security Agency can continue to collect the phone records of millions of Americans, but only for three more months, under a ruling Friday by the U.S. appeals court here. (Savage 2015)

"Justice Department: Agencies Need Warrants to Use Cellphone Trackers." September 4, 2015.
The Justice Department announced a policy Thursday that will require its law enforcement agencies to obtain a warrant to deploy cellphone-tracking devices in criminal investigations and inform judges when they plan to use them. (Nakashima, "Justice Department" 2015)

Many of the interactions concerning the legality of these data collection programs take place in public forums in what appear to be transparent ways. That is, we see legislative action, court rulings, and press coverage of the ongoing controversies. In some ways, it appears that democracy is alive and well and that new media are playing an important role in communication interactions among stakeholders, including everyday citizens. The problem, of course, is that the status quo features an enormous number of factors that hide data collection, surveillance, and data analytics in plain sight.

The Foreign Intelligence Surveillance Court (FISA court) is tasked with overseeing requests for surveillance warrants related to investigations of potential terrorist activities. The fact that the FISA court functions as a kind of check and balance against illicit surveillance and data collection is supposed to reassure all parties—politicians, intelligence operatives, and the public alike—that information-gathering activities are under control. At various times, small amounts of information about the functions of the court have been released or leaked. Generally, though, the court works in secret. For example, it has been reported that there have been entire years (2010 and 2011) during which the FISA court approved every single request it received for data collection activities (Farivar, "Secret Court" 2013). Although the court and its apologists defend its work (Ackerman 2013), secret government proceedings tend not to be the best way to reassure skeptics about the legality and benefits of other secret government procedures. Further, one cannot verify or validate the degree to which the court is scrutinizing intelligence activities.

With regard to the failure of certainty about constitutional protections under the law, one additional example will suffice. A recent US circuit appeals court ruled that the NSA could restart and continue its data collection programs that had been ruled illegal and barred by an earlier district court ruling (D.G. Savage 2015). However, the August 2015 circuit appeals court ruling applied only to three months of action, prior to the expiration of the authorizing law—and in the meantime, Congress passed legislation that continues parts of the program. Certain questions then arise: What does the NSA do with material that was collected before the order to stop? (Hint: They go on analyzing it.) How would anyone know whether the NSA followed the order that restricted the collection of data? (Hint: For the most part, we don't know.) After the order that enabled resumption of the program, did the NSA attempt to retrieve any of the data from the three-month lacuna and, if so, would that effort be illegal because the three months had been blocked out or be legal because the NSA had been told it is legal to collect data? (Hint: They probably did fill in the data and only higher powers know if that was legal.)

As noted by Robert McChesney, our national security complex is "almost unimaginable" and includes over 850,000 people with top-secret security clearances. Top-secret clearances are held by 1,300 government agencies and 2,000 private companies that collect intelligence data (2013: 161). According to McChesney, "It is a massive self-interested bureaucracy with no public accountability and barely a trace of congressional oversight." And what is the relationship between the government, the ISP cartel, and the digital giants? "The evidence suggests it is complementary and collegial, even intimate. ... The evidence is clear: The Internet corporations place a lower priority on human rights and the rule of law than they do on profits" (McChesney 2013: 162–163).

Secrecy prevails in both the private and governmental data marketplaces. In short, the machinations of our government's surveillance, data collection, and data analysis efforts are shrouded in secrecy, though sometimes in plain sight. The secrecy paradigm compromises fundamental principles of the FIPs.

· 4 ·
FIPs 2 AND 4
Discovery and Repair

> You can't fix what you can't see/It's the soul that needs the surgery.
> —"Pretty Hurts," Sia Furler, Joshua Coleman, and Beyoncé Knowles

> There must be a way for an individual to find out what information about him is in a record and how it is used.
> —FIP 2, *Records, Computers, and the Rights of Citizens*

> There must be a way for an individual to correct or amend a record of identifiable information about him.
> —FIP 4, *Records, Computers, and the Rights of Citizens*

A central assumption of this book, noted particularly in the introduction, is that the FIPs have been well known and available for a long time. Unfortunately, most of the players in the data marketplace ignore the principles. Likewise, a fully articulated set of principles and protocols exists with regard to managing some of the most sensitive information in our socioeconomic system: financial data. If the data marketplace were simply brought into line with the Fair Credit Reporting Act (FCRA) and the Fair and Accurate Credit Transactions Act (FACTA), many of the principles of the FIPs would be satisfied and a large percentage of privacy abuses would be ameliorated.

The FCRA dates from 1970; FACTA went into law in December 2003. The FCRA provides citizen-consumers with the following rights:

- You must be told if information in your file has been used against you.
- You have the right to know what is in your file.
- You have the right to ask for a credit score.
- You have the right to dispute incomplete or inaccurate information.
- Consumer reporting agencies must correct or delete inaccurate, incomplete, or unverifiable information.
- Consumer reporting agencies may not report outdated negative information.
- Access to your file is limited.
- You must give your consent for reports to be provided to employers.
- You may limit "prescreened" offers of credit.
- You may seek damages from violators.
- Identity theft victims and active duty military personnel have additional rights. ("A Summary of Your Rights" n.d.)

FACTA is "[a]n Act to amend the Fair Credit Reporting Act, to prevent identity theft, improve resolution of consumer disputes, improve the accuracy of consumer records, make improvements in the use of, and consumer access to, credit information, and for other purposes" (FACTA 2003). The act includes:

TITLE I—IDENTITY THEFT PREVENTION AND CREDIT HISTORY RESTORATION
Subtitle A—Identity Theft Prevention
Subtitle B—Protection and Restoration of Identity Theft Victim Credit History
TITLE II—IMPROVEMENTS IN USE OF AND CONSUMER ACCESS TO CREDIT INFORMATION
Sec. 211. Free consumer reports.
Sec. 212. Disclosure of credit scores.
Sec. 213. Enhanced disclosure of the means available to opt out of prescreened lists.
Sec. 214. Affiliate sharing.
Sec. 215. Study of effects of credit scores and credit-based insurance scores on availability and affordability of financial products.
Sec. 216. Disposal of consumer report information and records.
Sec. 217. Requirement to disclose communications to a consumer reporting agency.
(FACTA 2003)

Both the FCRA and FACTA map very closely on the FIPs. Although the two acts apply to data that is similar to information that appears in the data marketplace—and in many cases is the same data—the federal laws apply to only specific types of entities in the financial system, for example, banks, credit unions, and credit reporting agencies. It is remarkable to fathom how entities functioning within the data marketplace have been held beyond the reach of these laws. The FIPs were recommendations of a single governmental agency while the FCRA and FACTA are federal law. Merely applying these laws to the data marketplace would go a long way toward protecting citizens' privacy.

Unfortunately, as noted by David Lazarus in the *Los Angeles Times*, in the United States, business interests come before protection of citizens' personal information; data privacy is treated as a privilege rather than a right. While other countries, such as those in the European Union, begin privacy discussions focused on human rights, the US conversation begins with questions about how actions move markets and profits. In the United States, we have not only *not* regulated the Internet, we have not regulated most of the data marketplace. Elsewhere, among EU nations, for example, "Europe's nailed it: People have a right to privacy and businesses must honor that right" (Lazarus 2015).

This chapter focuses on two of the FIPs that are strongly supported, in contexts other than the data marketplace, by federal statutes like the FCRA and FACTA. Citizens should have the ability to know what personal information (of theirs) is contained within databases and they should be able to correct, and in some cases edit, those materials. This chapter begins by examining the arduous tasks that consumers face when attempting to learn about entities that are collecting and holding their personal information. The chapter then briefly reviews efforts from outside the United States to legislate increased online privacy, focusing especially on "the right to be forgotten," and relationships among these efforts and the ability of consumers to correct their records. The third section of the chapter investigates employer-employee relationships regarding negotiations over the ownership of social media accounts and account data—struggles that can result in further loss of control over personal data.

You Can't Fix It If You Don't Know About—And Can't Access—It

Learning about how, and how much of, our personal data is collected and transacted in the data marketplace is arduous, shocking, and daunting. In fact, the degrees of difficulty serve as pillars in the support structure of the status quo. The extent of the difficulties also illustrate how illegitimate our systems and circumstances have become. Change proposals found in chapter 7 notwithstanding, this book does not serve as a thorough instruction manual for how to protect your privacy. Still, in order for people to evaluate their personal risk models on the way to deciding how they want to treat their privacy and the data marketplace, they need to learn about the composition of, and uses for, their digital dossiers. Further, if citizens want to force enterprises to adhere to the second and fourth FIPs, they first need to determine which enterprises to target.

Writing in the *New York Times*, Natasha Singer laments:

> Our mobile carriers know our locations: where our phones travel during working hours and leisure time, where they reside overnight when we sleep. Verizon Wireless even sells demographic profiles of customer groups—including ZIP codes for where they "live, work, shop and more"—to marketers. But when I called my wireless providers … in search of data on my comings and goings, call-center agents told me that their companies didn't share customers' own location logs with them without a subpoena. Consolidated Edison monitors my household's energy consumption and provides a chart of monthly utility use. But when I sought more granular information, so I could learn which of my recharging devices gobbles up the most electricity, I found that Con Ed doesn't automatically provide customers with data about hourly or even daily use. … Then there is my health club, which keeps track of my visits through swipes of my membership card. Yet when I recently asked for an online log of those visits, I was offered a one-time printout for the year—if I were willing to wait a half-hour. … In our day-to-day lives, many of us are being kept in the data dark. (Singer 2013)

Angwin (*Dragnet* 2014) and Bazzell (2013) make very clear that finding out about one's digital dossier is arduous, difficult, and often frustrating work. There are vast numbers of players in the marketplace and grave difficulties in identifying, reaching, and influencing their data collection practices.

The initial points of contact for a large portion of data collection functions are the browsers used to access the Internet. Browser plug-ins are available that identify many of the enterprises that place cookies, use trackers, and make requests for data during Internet browsing sessions. For example, the Electronic Freedom Foundation sponsors the development and deployment of Privacy Badger, a browser plug-in that blocks trackers and tracking requests and identifies the enterprises behind those data collection efforts. One can also use the plug-in to learn more about those enterprises. A similar browser plug-in, Disconnect/Blur, enables users to block trackers and to protect their search results as private. DoNotTrackMe blocks trackers and can be set to block advertising. Numerous other products exist, some charging subscription rates, others offering free services, and yet others offering multiple product levels from 'freemium' to expensive subscriptions.

Users quickly learn that blocking trackers, cookies, and information requests is complicated technical work that often requires a level of understanding about digital systems and products beyond that of most average consumers. Further, examining the results generated from even simple and free-to-use browser plug-ins reminds users of the large numbers of entities engaged in the data marketplace. For example, a recent visit to the *New York Times* homepage using Chrome resulted in a Disconnect report of 15 data requests (3 from advertising outfits, 5 from analytics outfits, 1 from a content provider and 6 from Google or Facebook social media aspects). Disconnect/Blur blocked 7 different trackers. A stop at the homepage of my local newspaper, the Peoria *Journal Star*, found a Disconnect report of 42 data requests (4 advertising, 5 analytics, 20 content, and 12 social media); at this site, Disconnect/Blur blocked 9 trackers. Attempting to acquire reports of the data collected by this plethora of data merchants would be a daunting task.

Efforts to learn about one's digital dossier immediately direct attention to Google's many products and services. Angwin and Bazzell report varied levels of success from interactions with Google. A later section in this chapter discusses the right to be forgotten, a concept and set of practices that turns our attention, primarily, to Google's role in data collection and management. For now, we can note that Google has a product or service that touches almost every corner of the Internet; if Google doesn't touch it, its partners do. For instance, using information provided by the Google Dashboard (buried in her Gmail account settings) Angwin found that, over approximately six years, she sent email to 2,192 people, across 23,397 email and chat sessions. Further, she was able to learn that she averages approximately 26,000 Google searches each month (*Dragnet* 2014: 81). To discover much of anything about your digital dossier, it is fairly crucial to gain access to what Google knows about you. Unfortunately, going beyond the counts representing usage data that one finds within the dashboard is a very difficult process, as Google is relatively unresponsive to requests from users in the United States.

A lot of the material in our digital dossiers is based on our involvement with social media platforms. Users have a variety of goals for managing social media platforms. In some cases, they may want to stop using a given social media service. They may want to go a step further and remove their data from a service they have left. On the other hand, users sometimes want to continue participating in social media but with increased levels of control over the data that is shared by a given platform. Learning to use the software platform, and its various controls, is important. But sometimes, users need to interact with developers and operators in order to enact constraints or ascertain the various data collection functionalities found on a site.

Social networks are one of the Internet's primary privacy poison pills. Materials that users place within social networks are exceedingly important to the analytical work being done to produce targeted marketing and advertising. Users wanting to protect their privacy while continuing to use social media learn very quickly that it is far easier for them to place data on a social network than it is to find out what happens to that material after its placement or to constrain the uses to which it is put. Worse, fully removing materials from the Internet, and therefore the data marketplace, once the material has appeared on social networks is very difficult—perhaps impossible. Network effects spread information across users very quickly; removing information from one site seldom removes it from multiple sites. In every place the data appears, it becomes more fodder for collection and analysis by participants in the data marketplace.

Operators of some social networks are more cooperative than others when asked for information about user data. For example, Angwin (*Dragnet* 2014: 83) reports that Twitter quickly provided a spreadsheet detailing her contributions to that social media platform. As we will see in the "right to be forgotten" section, Facebook is significantly less forthcoming.

Although a lot of time and effort can go into examining the plethora of entities that collect personal data via basic functions like search, Web browsing, and social media use, the really hard work begins when one approaches the outfits that make up the great bulk of the data marketplace: data brokers and marketers. Bazzell (2013) identifies two major types of data brokers: those that focus on public data and those that focus on nonpublic data. Bazzell lists 10 primary public data brokers, including Acxiom, LexisNexis, Intelius, Rapleaf, and Merlin Information Services. Angwin (*Dragnet* 2014) contacted 23 of the data brokers that she had identified as attending to her Web browsing, and she was able to acquire some information about her data from 13 of those. However, she was generally unable to acquire significant amounts of data within her dossier through data brokers. Bazzell also provides information about a related set of data brokers: the data marketers. These enterprises work in a more business-to-business fashion than do general data brokers. Data marketers attempt to match the sales interests of a given business with specific appropriate audiences. These outfits include Epsilon, Valpack, DirectMail, and DMA Choice, among many others.

Angwin identified another segment of the data marketplace: the data-scoring business. In data scoring, data points are turned into data sets, and then data sets are turned into individual profiles with scores for various functions that appeal to potential marketing and advertising outfits. Data-scoring outfits produce the kinds of categories that, Turow reminds us, eventually label individuals as advertising and marketing "targets or waste" (Turow 2011: 88). Data scoring is particularly relevant within an additional network of data collection and analysis enterprises: advertising exchanges. Advertising exchanges function as the new digital analog to older advertising agencies and advertising network models by pooling and modifying information about users toward developing targeted advertising and marketing strategies. For example, "eXelate says it gathers online data from two hundred million unique individuals per month through deals with hundreds of websites that see an opportunity for revenue based on the firm's visitor analysis. ... BlueKai, an eXelate competitor, also strikes deals with thousands of websites using cookies to collect anonymous data about visitors' activities from that information to make inferences about the kinds of purchases that might interest those people" (Turow 2011: 79–80).

While much of the information within the data marketplace is strongly related to delivering targeted advertising to specific users, additional segments play roles that support targeted marketing but that also provide data to interested parties outside of marketing-focused businesses. For example, Bazzell notes the importance of getting control over the information provided to online directories of people (2013). These sites collect information from publicly available sources, such as tax data, social networks, resumes, marketing databases, and various other database outlets. Spokeo, Radaris, and 123 People are examples. People directories also utilize information found in more traditional telephone directory outlets. White page and yellow page services—along with a long list of related business endeavors—catalog, print, deliver, and in many cases sell basic information about names, phone numbers, and addresses. Efforts to de-anonymize data sets very often require and use basic information like home addresses. Consumers wanting to maintain control over their privacy need to address the role of people directories. The public nature of the information repurposed by these outfits makes controlling them very challenging.

Public records are major sources of information for the data marketplace and present a conundrum for advocates of open government, open records, and privacy. The old saying 'nothing disinfects better than sunlight' captures an important point of advocacy for open government and open records supporters. Generally speaking, if and when the government has collected information, according to the FIPs, the citizens who have provided it should be able to access it and correct it. These two aspects are, after all, the point of this chapter.

In various ways and across a wide variety of agencies, governments have used electronic databases and networks to give citizens access to information collected by the government (other than intelligence and law enforcement data). This seems to be as it should be; after all, information derived from the people belongs to the people. Unfortunately, it also exemplifies the saying 'be careful what you wish for—you just might get it.' Users can examine, and sometimes correct, the data in their own publicly available records. However, these records also become available to marketers and advertisers (and others, for example, law enforcement officials) wanting to leverage the information within the data marketplace (or for other uses). Property tax records, property sales records, voter registration records, military recruiting databases, date-of-birth records, ancestry archives, court records, and driver's license facility records are among those public records that are useful to citizens but difficult to control.

Further, and more dangerously for privacy, as noted in chapter 1, public information is sometimes leveraged as a way to gain seemingly unfair advantages in political controversies. For example, supporters of Republican Governor Scott Walker of Wisconsin accessed the email records of professors in the University of Wisconsin system as a way to identify faculty members who had indicated via email their opposition to the governor's anti-union policies and sentiments. Contents of those emails were then published and used in campaigns against the governor's opponents (Finnegan 2011). Although professors at public universities are in fact public employees, and as such their email messages become public documents, the wisdom of allowing these materials to be accessed by third parties and leveraged in political controversies is questionable.

Two additional areas of interest will be mentioned before we examine the right to be forgotten: the potential for data abuse by ISPs and the role of mobile phone providers.

In effect, (potentially) the most efficient pieces of surveillance technology in our homes and offices are the devices that connect our computers and Wi-Fi-enabled devices to the Internet. Modems, routers, and the servers that Ethernet cables connect provide access to, and records of, every aspect of our interactions over the Internet. In many cases, the entities functioning as our ISPs do not monitor or collect content-specific information about our online behaviors. The status quo is not guaranteed to last forever, nor is it foolproof in every case even in the present. This is one of the worrisome aspects of industry conglomeration and market dominance in the ISP business sector. If the dominant player, let's say Comcast, were to decide that the devices connecting users to the Internet should be two-way receivers and broadcasters, Comcast would be able to access and record information about not only our television viewing habits but all of our Internet behaviors. That would provide a single entity with an enormous amount of very valuable personal data.

Regulatory maneuvering by the Federal Communications Commission (FCC) resulted in the establishment of rules supporting "net neutrality" ("FCC Adopts" 2015). These regulations followed a series of court battles leading to action on the part of the FCC to define the Internet as a "telecommunications service," thereby enabling the FCC to regulate aspects of the US Internet infrastructure ("Protecting and Promoting" 2015). On March 10, 2016, the FCC issued the "Broadband Consumer Privacy Proposal Fact Sheet," detailing proposals "that would significantly curb the ability of companies like Comcast and Verizon to share data about their customers' online activities with advertisers without permission from users" (Kang, "FCC Proposes" 2016). The FCC's actions constitute the government's most direct regulation of privacy in the commercial environment. The FTC's treatment of privacy policies and ToS could further strengthen governmental oversight and enforcement, especially if that agency begins to treat wrap contracts as unfair business practices.

This chapter will not engage in an extended discussion of mobile phone providers. Still, the Snowden revelations and other incidents make clear that mobile phone service providers play significant roles in the security and privacy equation. The symbiotic relationships among governmental intelligence operatives, law enforcement agencies, and telecommunications providers entail crossing and mixing private and public data streams in ways that compromise privacy under the guise of increasing security. Further, market realities clearly indicate significant shifts toward the importance of mobile phones, in contrast to the rapidly dwindling role of wired devices as well as in terms of the increasing dominance of mobile access to the Internet.

In the context of the five FIPs, the most efficient place for us to take up questions about mobile phones is in chapter 6, during examinations of database managers' responsibilities for protecting private data.

Online Privacy and the European Union

Regulators and courts throughout the EU have long taken a view that is more protective of citizens' privacy rights than the US perspective. American technology companies, and the federal government, have long resisted efforts on the part of European regulators to enable their citizens to better protect personal privacy. Long-standing philosophical differences have recently coalesced into battles over specific policy proposals. For example, in 2013, tech companies in Silicon Valley and the federal government both pushed back against "Europe's effort to enact sweeping privacy protection for digital data":

> Several proposed laws working their way through the European Parliament could give 500 million consumers the ability to block or limit many forms of online Web tracking and targeted advertising. All the major American tech companies have directed their lobbyists in Brussels, where the Parliament is based, to press to weaken or remove these proposals from the European provisions. (O'Brien 2013)

Google stands as both the starting place and the target for regulatory actions by the EU. Clearly, Google's dominance in search, and its prominent roles in the data marketplace, encourages regulators to test proposals and apply policy recommendations against the Google model first.

For example, in September 2014, EU privacy regulators issued a series of guidelines for various ways in which Google should change its privacy policies and protocols, especially within search functions (Essers 2014). These recommendations included a variety of procedures, policies, and functionalities that would significantly change both the ways that Google's base software operates and the degree of control users have over their personal data. The recommendations suggested that privacy policies be easy to reach, read, and understand. Modalities for customizing policies to the privacy laws in particular jurisdictions, and for specific customers, were proposed. Users were to be given ways to ascertain what data is being held as well as to correct it if necessary. Google was also asked to clearly identify entities with which it shared data rather than merely referring to "our partners." European regulators also specified that they be given opportunities to review Google's data retention policies and practices (Art. 29 WP 2014).

Various EU courts have indicated that the 'business as usual' practices of US technology companies—especially Google and Facebook but also other companies—face judicial challenges in addition to regulatory constraints. For example, a Spanish court ruled in 2014 that an Internet search engine operator is responsible for, and may be sued over, links that provide access to personal information found on third-party pages ("Judgment in Case C-131/12" 2014). This case—*Google Spain v. Agencia Española de Protección de Datos (AEPD) and Mario Costeja González*—serves as the rulemaking precedent establishing the right to be forgotten that is being used to force Internet search engines (particularly, and initially, Google) to remove links that contain incomplete or inaccurate information. The ruling also establishes authority behind requests for data records, since those are necessary in order to make valid requests for deleting links. In 2015, a UK court of appeal confirmed an earlier high court decision that a group of British consumers (using Apple Corporation's Safari browser and Google's various services) can pursue privacy litigation in the United States rather than only in the UK (*Vidal-Hall & Ors v. Google* 2014). The litigation was brought in response to consumers' complaints that their personal information is being leveraged in the data marketplace in support of targeted advertising (Moody 2015). Both of these rulings, if upheld and extended to additional providers and new media enterprises, would severely challenge the status quo in the data marketplaces while supporting increased privacy protection for users.

This trend continued in a significant way in 2015 as the European Court of Justice struck down a major data-sharing pact between the United States and Europe, "finding that European data is not sufficiently protected in the United States":

> The ruling will affect more than 4,400 U.S. and European companies that rely on the agreement to move data back and forth across the Atlantic to support trade and jobs. It also could have huge implications for U.S. intelligence agencies, which depend on an ability to sift through large volumes of data in search of clues to disrupt terrorist plots. The decision invalidated the Safe Harbor framework of 2000, reached between the United States and the European Commission. ... [The] ruling grew out of revelations by a former National Security Agency contractor, Edward Snowden, about the scope of NSA surveillance. (Nakashima, "Top EU Court" 2015)

The ruling and the resultant policy changes were driven at least in part by the actions of a private citizen living in Vienna, law student Max Schrems. After spending time as a visiting student at Santa Clara University in the Silicon Valley, Schrems became alarmed by the fact that US companies, doing business in Europe, were not following EU privacy laws. In 2011, Schrems leveraged the European "right to access" laws and made requests for his data records from Facebook. After receiving voluminous reports, he filed 23 claims against Facebook's European office in Ireland (Angwin, *Dragnet* 2014: 82–83; Fioretti, Nasralla, and Murphy 2015). His last claim worked its way to the EU high court, resulting in legislative action barring the in-place data exchange practices (between the United States and the EU) of new media companies. Sometimes the actions of a determined individual can make a difference, even against giant multinational corporations and industries.

In order to fulfill the goals of the fourth FIP (to be able to correct personal records), one must be able to attain the goals of the second FIP (to ascertain the contents of those records). As noted, legislation and regulation in the United States requires that credit agencies provide annual reports to citizens, without charge, on proper request. No such modality exists within the data marketplace. As illustrated by the Schrems case, obtaining data records is a crucial first step that is supported in the EU by a government regulation. As exemplified by Angwin (*Dragnet* 2014), Bazzell (2013), Singer (2013), and others who have tried to obtain their records, this first step is nearly impossible to accomplish in the United States, where no regulations compel enterprises in the data marketplace to comply with individuals' requests.

Although many of the judicial and legislative actions taken by EU regulators focus on broad issues related to human rights and privacy, the practice that is gaining the most traction is the somewhat narrower requirement that Google expunge inaccurate or otherwise objectionable material at the request of users. Administration of the right to be forgotten has become the central battleground for privacy advocates; successes in the EU seem to indicate that US technology companies can be reined in, if and when governmental regulators act in support of the personal privacy of the constituents.

The Right to Be Forgotten

The Spanish court's ruling in *Google Spain v. AEPD and Mario Costeja González*, and threats of further penalties by EU regulators, convinced Google to design procedures whereby Europeans could make "right to be forgotten" requests. Subsequent developments shed light on what can happen when Internet regulations bump up against the interests of multinational corporations.

In 2014, Google reported having received about half a million requests between May and October that links to data be removed; the company claimed to have removed approximately 42% of those links (Kravets, "Google Has Removed" 2014). Across the first year of the new procedures, Google removed approximately 59% (322,000 of just over 500,000) (Kravets, "Google Rejecting"

2013). Additional, related actions soon followed: In August 2015, Google was ordered to remove links to stories that reported on links that had been removed as a result of "right to be forgotten" requests (Miners 2015). By September 2015, courts in France had ruled that Google is required to remove "right to be forgotten" links from all of its systems rather than just from systems within the country where the request originated.

As could be expected, efforts to extend the right to be forgotten to the United States have taken hold. In July 2015, the group Consumer Watchdog enacted a new strategy designed to align US law with European practice. The group appealed to the FTC, arguing that Internet industry protocols that prevent Americans from receiving data reports and enacting "the right to be forgotten" are unfair and deceptive trade practices.

The costs and benefits of adopting similar systems in the United States are hotly contested (Quinn and Brachmann 2015). Suffice to say, the European actions have changed the game and destabilized the status quo, particularly abroad but with long-term consequences for new media enterprises on both sides of the Atlantic. Nations of the EU have shown that governments, courts, and regulatory agencies can change the ways that new media giants do business. Directing American new media companies toward practices that support the FIPs is possible and may be preferable to the status quo found in the United States today.

Ownership of Social Media Accounts in the Enterprise: "Unhand My Stuff!"

Client lists have long been treated as protectable intellectual property under US trade secret laws. Recent cases illustrate the emerging importance of "ownership" over social media accounts, with regard to both trade secrets and privacy questions. For example, *PhoneDog LLC v. Kravitz* (2011) indicates that employers can rightfully claim ownership of social media accounts, while *Christou v. Beatport, LLC* (2012) and *Amway Global v. Woodward* (2010) suggest that employees can sometimes maintain control over accounts and passwords. While companies have valid interests in protecting company information, especially trade secrets, employees often use social media, sometimes at work and/or for employers, in ways that strike them as personally 'private.' When companies take over accounts that employees had treated as personal, data that employees had seen as private is treated as non-private, sometimes being circulated far more broadly than the employees had intended.

A number of states have taken up the issue of privacy and ownership over social media accounts. The resultant legislation is somewhat surprising in that it tends to protect employees more than employers—a pattern that is the opposite of what one finds in trade secret law. Although the statutes are not exactly the same, they share numerous similarities. In general, they constrain employers from forcing employees to disclose passwords and content of their social media accounts (Rubin and Stait 2013; Milligan, "California" 2013, "Michigan" 2013; Schaefers, "Nevada and Colorado" 2013, "Washington State" 2013). However, the laws contain nuances and various exceptions. Employees may be hired for the purpose of, or charged with the task of, developing or managing a business's social media presence and accounts. When the resulting social media account is a 'work for hire,' the employee cedes all rights to the employer. Likewise, the law passed in 2013 in the state of Nevada constrains employees from suing employers for violations of the law, limiting them instead to filing complaints with the appropriate state agency.

It is clear that contestation over the ownership of data within social media accounts in the workplace will continue ("Fact Sheet 7" 2016). As noted in a 2014 hearing before the US Equal Employment Opportunity Commission (EEOC):

> The use of social media has become pervasive in today's workplace and, as a result, is having an impact on the enforcement of federal laws. ... Jonathan Segal, speaking on behalf of the Society for Human Resource Management (SHRM), explained that employers use different types of social media for several different reasons: employee engagement and knowledge-sharing ...; marketing to clients, potential customers and crisis management; and for recruitment and hiring of new employees. In fact, ... 77 percent of companies surveyed reported in 2013 that they used social networking sites to recruit candidates, up from 34 percent in 2008. ... In one reported decision arising from the federal sector, EEOC's Office of Federal Operations found that a claim of racial harassment due to a co-worker's Facebook postings could go forward. Additionally, in response to a letter from Senators Charles Schumer and Richard Blumenthal, the EEOC reiterated its long-standing position that personal information—such as that gleaned from social media postings—may not be used to make employment decisions on prohibited bases, such as race, gender, national origin, color, religion, age, disability or genetic information. ("Social Media Is Part" 2014)

Access by employers and potential employers to the personal and private information found within corporate and individual social media accounts raises a number of concerns and aspects surrounding privacy. Determining ownership of the materials, as well as establishing the degree to which the materials qualify as 'private,' adds complexities to questions about the personal data that people post to their privately owned and curated social media accounts. Questioning how individuals can claim to want privacy when they post intimate details to online social media accounts carries over to the work environment, when information is posted under the auspices of business practices. On the other hand, gainful employment should not trump personal rights; often, employees want to discuss work-related issues on social media while employers might want to quash all discussion of business matters in communication venues outside of corporate control. To this point, state legislatures appear to recognize and appreciate workers' rights and the (sometimes unfair amount of) leverage that employers hold.

Toward Chapter 5 and the Third FIP

In many ways, the third FIP might well be the most important. The third FIP raises pragmatic questions about many of the privacy problems in the commercial data marketplace. It might also suggest the most straightforward solutions to these privacy issues. In the commercial realm, regardless of whether one knows about a given data collection practice (the first FIP), and whether one can learn what information has been collected and/or make changes to it (the second and fourth FIPs), using data for the specific purpose(s) for which they were given/taken (the third FIP), *might* pass muster. Chapter 5 examines these and related aspects.

· 5 ·
FIP 3

One Use Should Not Bind Them All

> I gave you all/And you rip it from my hands/And you swear it's all gone.
> —"I Gave You All," Marcus O. J. Mumford, Edward J. M. Dwane, Benjamin W. D. Lovett, and Winston A. A. Marshall

> There must be a way for an individual to prevent information about him obtained for one purpose from being used or made available for other purposes without his consent.
> —FIP 3, *Records, Computers, and the Rights of Citizens*

Although each of the FIPs is important, and arguments can be made in favor of any one of the five being the most critical, the third FIP seems to be the most attuned to the views of typical US citizen-users. The full title of a 2016 report from the Pew Research Center captures the sense of it: "Privacy and Information Sharing: Many Americans Say They Might Provide Personal Information, Depending on the Deal Being Offered and How Much Risk They Face" (Rainie and Duggan 2016). Americans' everyday media consumption and ubiquitous digital use attest to their powerful attraction to new media products and services. On the face of it, there seems little reason to suspect that Americans retain any concern for privacy or the uses to which their personal data are put.

Yet, results of survey research appear to indicate that Americans are very concerned over the uses to which their personal data are put. In fact, their responses show how important the third FIP could be to their cost-benefit analyses for digital participation. I write "could be" because the secrecy surrounding the collection and leveraging of personal data hinders a thoroughgoing cost-benefit analysis.

In common language, US citizens are willing to—are perhaps even eager to—exchange personal data with providers of services/products. This impulse certainly holds true within environments where the service/product appears to be free, or of exceedingly low cost, with the exchange of personal data as the apparent price to pay for the service/product. Americans understand that information does not actually 'want to be free' and that giving personal data to vendors and providers constitutes a form of payment—payment that users generally approve of in exchange for specific products/services.

However, US citizens are concerned when they suspect that their personal data is used for additional or other purposes. They are also concerned about cases in which they are unaware of, or unable to ascertain, the uses to which their data are being put. Rainie and Duggan specify some of the issues in their summary of findings, noting that such terms as "'creepy' and 'Big Brother' and 'stalking' were used regularly in the answers of those who worry about their personal information" (2016: 5):

> They [respondents] also regularly expressed anger about the barrage of unsolicited emails, phone calls, customized ads or other contacts that inevitably arises when they elect to share some information about themselves. ... The initial bargain might be fine, but the follow-up by companies that collect the data can be annoying and unwanted. ... Some of the most strongly negative reactions came in response to scenarios involving the sharing of personal location data. Many Americans express suspicion that data collectors (from employers to advertisers) have ulterior motives in their pursuit of personal data. (Rainie and Duggan 2016: 3, 5, 6)

This 2016 survey does not represent the first time that Americans had indicated concerns over secondary uses of their personal data. Its findings are consistent with research that followed the revelations by Edward Snowden. Shifts in public opinion are sensitive to historical conditions. For example, in the post-9/11 era, particularly after the bombings at the Boston Marathon, Americans expressed almost enthusiastic willingness for additional video surveillance cameras in urban areas, regardless of their lack of knowledge about who was collecting the data or to what uses the data would be put (Handler and Sussman 2013).

After the Snowden revelations, public opinion moved away from acceptance of unlimited collection and use of personal data. In the report of a study published in November 2014, Mary Madden noted a number of circumstances in which Americans object to secondary and third-party uses of personal data:

> 80% of those who use social networking sites say they are concerned about third parties like advertisers or businesses accessing the data they share on these sites.
>
> 70% of social networking site users say that they are at least somewhat concerned about the government accessing some of the information they share on social networking sites without their knowledge.
>
> 80% of adults "agree" or "strongly agree" that Americans should be concerned about the government's monitoring of phone calls and internet communications. Just 18% "disagree" or "strongly disagree" with that notion.
>
> 81% feel "not very" or "not at all secure" using social media sites when they want to share private information with another trusted person or organization.
>
> 61% of adults "disagree" or "strongly disagree" with the statement: "I appreciate that online services are more efficient because of the increased access they have to my personal data." (Madden 2014: 3)

As noted in chapter 1, Americans did not automatically and easily accommodate themselves to giving up personal information in exchange for online products and services. Privacy concerns were often listed as reasons for why consumers were initially unwilling to participate in the 'interactive' new media marketplace (Catacchio 2010).

As Web 2.0 progressed (after 2005), online enterprises cultivated consumer acceptance of trading data for services. By 2010, many online retailers had learned to leverage Americans' increasing cooperation with data collection tied to online commerce:

> The budgeting Web site Mint.com, for example, displays discount offers from cable companies or banks to users who reveal their personal financial data, including bank and credit card information. The clothing retailer Bluefly could send offers for sunglasses to consumers who disclose that they just bought a swimsuit. And location-based services like Foursquare and Gowalla ask users to volunteer their location in return for rewards like discounts on Pepsi drinks or Starbucks coffee. These early efforts are predicated on a shift in the relationship between consumer

and company. Influenced by consumers' willingness to trade data online, the sites are pushing to see how much information people will turn over. (Clifford 2010)

A high point for heralding location data as 'the next big thing' came in 2010. For example, O'Reilly's "Where 2.0" conference that year "[brought] together the people, projects, and issues building the new technological foundations and creating value in the location industry" ("The 2010 Where 2.0" n.d.) and Convergence Labs' "Geo-Loco Conference" promised to present "the next big thing in advertising, social media and discovery!" ("Geo-Loco" 2010)

By 2012, online behavioral tracking had increased by as much as 400% (Farivar, "Online Behavioral Tracking" 2012). Users showed a developing preference for conveniences over privacy. For instance, a survey of 4,000 people indicated that users appear to be aware of the privacy risks of third-party log-ins—46% knew that their data would be sold—yet 60% of respondents still wanted that convenience (Hsukayama 2014).

As the data marketplace grew, the multiplicity of uses for collected data also increased. Tracking consumers, and using their real names, became a hot trend among new media enterprises. "Previously, data brokers primarily sold this data to marketers who sent direct mail—aka 'junk mail'—to your home. Now, they have found a new market: online marketing that can be targeted as precisely as junk mail. … The marriage of online and offline is the ad targeting of the last 10 years on steroids" (Angwin, "Why Online" 2014).

Citizens wanting to use services/products in the new media environment are faced with daunting challenges when making decisions. On the one hand, they want the conveniences provided, yet they are at least marginally aware that their private data is being collected in exchange. Control of the collection and usage of that data is either so difficult or so far out of conscious reach that most consumers spend little time considering how they might attain that control. The situation presents a challenging and complex Catch-22: The data are put to so many uses, and those uses are often so very far away from the intention that the citizen/consumer had when offering them in trade, that participating in the system at all seems perilous and yet is ubiquitous and commonplace.

As the IoT comes to the forefront, many everyday objects will be connected to networks, collecting and sharing information about behaviors and activities. Some items will be 'wearables' that perform a variety of monitoring, reporting, and recording tasks. For example, personalized fitness monitors will help people leverage best practices in diet and exercise as well as check for optimal bodily functioning; the individual is thereby aided. Additionally, society as a whole can benefit when "the aggregate data being gathered by millions of personal tracking devices ... [indicates] patterns that may reveal what in the diet, exercise regimen and environment contributes to disease." Besides tracking the health of individuals, such data could perhaps also "inform decisions about where to place a public park and improve walkability" or "find cancer clusters or contaminated waterways" (Cha 2015). And yet:

> Federal patient privacy rules under the Health Insurance Portability and Accountability Act don't apply to most of the information the gadgets are tracking. Unless the data is being used by a physician to treat a patient, the companies that help track a person's information aren't bound by the same confidentiality, notification and security requirements as a doctor's office or hospital. That means the data could theoretically be made available for sale to marketers, released under subpoena in legal cases with fewer constraints—and eventually worth billions to private companies that might not make the huge data sets free and open to publicly funded researchers. (Cha 2015)

The pace and breadth of data collection, repurposing, and distribution have become so ubiquitous that both the letter and spirit of the third FIP are in full retrograde. Acxiom claims that it has information on nearly every US consumer, while Datalogix collects information from store loyalty cards across more than 1,400 brands and more than $1 trillion in consumer spending (Beckett 2014).

At a point, one may even have to question the motivations behind what appear to be socially responsible actions on the part of participants in the data marketplace with respect to collecting and repurposing information for profit. When Facebook offers free or low-cost Internet access to populations with high poverty rates, critics refute the company's claims that Internet access changes overall poverty rates. Perhaps there's more evidence that providing Internet access to people benefits companies that collect data about users then there is evidence in support of the notion that Internet access lifts poor people out of poverty:

> If Facebook can hook new Internet users early, it stands to gain from network effects, in which businesses with the most participants become exponentially more valuable. The effects are especially profound for digital businesses such as Facebook's. "Facebook shouldn't be talking about this as a charitable effort, but as a market development effort," said Ethan Zuckerman, director of the Center for Civic Media at MIT. ... Facebook's efforts have been so successful that in the Philippines, Thailand and Indonesia, more people report using Facebook than the Internet, apparently unaware that Facebook runs on the Internet. (McFarland 2016)

As useful as Apple's move to end-to-end encryption is, and in spite of protestations on the part of its leadership extolling its own support of civil liberties, the company has a long history of delivering products that have compromised user privacy either directly or indirectly, and of working with the government to unlock devices at the behest of law enforcement (Benner and Perlroth 2016; Goodin, "Contrary" 2013, "Researchers Find" 2015; Keizer 2011). Tech corporations that try to protect privacy are to be commended. Yet, corporate efforts that seek only to point the finger at, and perhaps temporarily limit, government intrusion seem self-serving and largely rhetorical. Claims made by government entities that they protect consumer privacy ring equally hollow, as "the clash continues between consumer advocates who warn that unfettered commercial data collection could chill the daily routines of Americans—where they go, what they say, how they shop—and industry advocates who warn that any restrictions on data collection could chill innovation" (Singer 2013).

Putting collected data to multiple uses without the direct knowledge of, and consent by, the consumers who provide the data presents myriad challenges to tech companies, government entities, the public, and privacy alike.

Data Collection and Security: Commercial Entities Sharing with the Government

Relations among various kinds of private information sharing, involving the government, that compromise single uses will be examined across four aspects: encryption and cryptography; export controls; the Snowden revelations; and crossing government and commercial data streams within the data marketplace.

Battles over Encryption and Cryptography

One of the principal ways to limit cross-purpose leveraging of private data is to encrypt content so that only the intended recipient can decode and use the information.

The US military and intelligence communities have worked diligently at cryptography, hoping to stay ahead of commercial and foreign entities in the development and deployment of encryption. The mission and goal of the NSA—leaders of the US government's cryptography efforts—is, in the agency's own words, "to gain a decision advantage for the Nation and our allies under all circumstances" (NSA 2016). The "Crypto Wars" between governmental and non-governmental workers (including proprietary companies, researchers, and independent hackers) are well documented by Steven Levy (2001), Kurt Opsahl (2015), and Kehl, Wilson, and Bankston (2015), among others.

The symbiotic relations among US military and intelligence operatives, researchers, and commercial developers have long been exceedingly tense, dramatic, confrontational, and varied. In general, high-tech and new media companies have been able to maintain relatively strong encryption standards domestically while having to dilute those standards in exported products and services. Despite the fact that various types of strong encryption are available domestically, few everyday Americans use encryption knowingly, often, or purposefully. Further, in some cases even the strong encryption that is available domestically fails to constrain data to single uses. Apparently, when the government wants to collect information about citizens, it is able to do so. The history of these relationships, in detail, far exceeds the scope of this volume. A brief review orients the discussion.

In general, governmental entities probably exceeded the technical capabilities of researchers and commercial entities during the early development of computers and global networks. Because of the secrecy involved in the spying business, it is difficult to establish an exact point when commercial interests and/or academic researchers caught up with or exceeded the capabilities of government-sponsored cryptographers. However, it is clear that commercial cryptography developed encryption capabilities during the 1990s that raised the level of concern among governmental entities. The results were, and are, an ongoing struggle over the degree to which government entities can control commercial encryption, either by limiting its strength or by decrypting code.

Governmental entities take the position that operatives, terrorists, and criminals—especially foreign but also domestic—take advantage of strong encryption and thereby put American lives, infrastructure, and property at risk. Military and intelligence agents and agencies want always to be able to decode the bad guys' communication and to be able to protect US secrets.

Commercial entities depend on encryption in order to enact digital rights management (DRM) for their products/services, especially within networked environments. Without encryption, commercial products and services lay open for rampant piracy. Even with DRM via encryption, piracy remains a serious threat to corporate profits and the economic viability of US commercial new media industries. Further, commercial entities want to convince customers that their consumer privacy is protected, thereby encouraging continued commercial vibrancy. Whether commercial interests actually do so is, of course, questionable—that is to say, the ways that commercial enterprises leverage private data within the data marketplace casts doubt on their claims about 'protecting user data.' Regardless, US commercial entities must maintain a rhetoric that is supportive of privacy. Most of the major commercial entities—for example, Google, Facebook, Apple, and Microsoft—have repeatedly made explicit statements about the relationships between their business successes and consumer trust. Commercial entities argue that if they cannot prevent the government from freely accessing customer data, consumers are less likely to trust the commercial entities and therefore will purchase fewer products/services.

The global nature of contemporary commerce requires that businesses are able to compete outside the United States. Export restrictions, often based on encryption standards, impact American business; these restrictions are examined in the next section. Here, we must acknowledge the interplay of government and commercial entities over these issues.

Researchers face the dilemma of wanting to fully understand and control the cutting edge of cryptographic knowledge. This goal challenges governmental entities. While commercial research and development of cryptography sometimes develops in proprietary secret, academic research in the field has long operated with academic freedom and open exchange of ideas. However, as the computer and new media ages matured, researchers in cryptography encountered increasing constraints by governmental entities. The US intelligence community places higher value on our abilities to keep secrets and crack enemy codes than in freely exchanging information about encryption, because such open exchanges threaten to empower 'our enemies' with code that our operatives can't break.

Tech companies in the United States receive an enormous number of data requests from governmental entities (intelligence, law enforcement, and the courts). Many of those requests are under Patriot Act (and other legal) protocols that keep the details secret. The government and the commercial operations have fought numerous battles, in and out of court, over making information about these requests (and corporate compliance) public. Reporting on secret requests is generally limited to listing the gross numbers of requests and compliance rates. These reports don't lift the veil of secrecy.

US commercial entities complain that the government makes far too many requests for private data. Privacy advocates decry governmental leverage of commercial data collections. The scope of the practice is enormous, almost beyond imagining. Even before Snowden's identity was revealed, articles detailing mass surveillance and data collection programs (such as PRISM) had appeared (especially in the *Guardian*, *Washington Post*, and *Baltimore Sun*). Using materials acquired from Snowden, Gellman and Poitras (2013) reported on data sharing among nine technology companies—Microsoft, Yahoo, Google, Facebook, PalTalk, AOL, Skype, YouTube, and Apple—and US intelligence. In 2013, the US Department of Justice obtained more than 21,000 orders for call records (Palazzolo 2014) and Verizon received over 321,000 legal orders for user data (Farivar, "Verizon Says" 2014). Further,

> Google received FISA requests related to the content of between 9,000 and 9,999 accounts during the first half of 2013. ... The court sought content related to 15,000 to 15,999 accounts from Microsoft. ... At Facebook, during the second half of 2012, the court sought data related to the content of 4,000 to 4,999 Facebook user accounts, the company said in a report. The company received more FISA requests during the first half of 2013—the number rose to between 5,000 and 5,999 users' accounts. (Miners 2014)

The trend continued into the first half of 2015, as Twitter reported receiving 2,871 requests and Apple reported answering approximately 7,100 of roughly 11,000 government requests (Isaac 2015, 2016). Apple's February 2016 refusal to write code the FBI could use to defeat the (standard iOS8) encryption on the iPhone of one of the San Bernardino murderers/terrorists may have seemed unexpectedly uncooperative (Apuzzo, "Should the Authorities" 2016; Benner and Perlroth 2016; Wingfield and Isaac 2016); however, by that time, many of the new media technology companies had spent years working toward redefining their positions in terms of the increasing numbers of government requests for private data.

In the first place, many small technology companies were increasingly unable to service the requests—so much so that a cottage industry of third-party providers sprang up to meet the demands:

> Not every Internet provider can handle the demands of a Foreign Intelligence Surveillance Act warrant or law enforcement subpoena for data. For those companies ... the answer is to turn to a shadowy class of companies known as "trusted third parties" to do the black bag work of complying with the demands of the feds. ... [S]maller ISPs who aren't frequently hit with warrants can't afford to keep the infrastructure or manpower on-hand to respond to requests—so they sign up with a "trusted third party" capable of doing the work as an insurance policy against such requests. Companies

such as Neustar, Yaana Technologies, and Subsentio contract with smaller providers and reap the profits from charging federal law enforcement and intelligence agencies for the data. Neustar and Yaana are also essentially private intelligence companies, providing large-scale data capture and analytics. (Gallagher, "When NSA" 2014)

In response to the increased demands and burgeoning tensions between the commercial and governmental entities, a number of leading new media corporations—including AOL, Dropbox, Evernote, Facebook, Google, LinkedIn, Microsoft, Yahoo, and Apple—banded together toward government surveillance reform. All in this group face a near-constant barrage of government-based data requests; many participated in numerous meetings with government officials. The alliance recognizes governments' needs to leverage information to protect their citizens' safety and security but calls on the governments to reform their information laws and practices. Recommended principles include:

1. Limiting Governments' Authority to Collect Users' Information
Governments should ... compel service providers to disclose user data that balance their need for the data in limited circumstances, users' reasonable privacy interests, and the impact on trust in the Internet. In addition, governments should limit surveillance to specific, known users for lawful purposes, and should not undertake bulk data collection of Internet communications.

2. Oversight and Accountability
Intelligence agencies seeking to collect or compel the production of information should do so under a clear legal framework in which executive powers are subject to strong checks and balances. Reviewing courts should be independent and include an adversarial process, and governments should allow important rulings of law to be made public in a timely manner.

3. Transparency about Government Demands
Governments should allow companies to publish the number and nature of government demands for user information. In addition, governments should also promptly disclose this data publicly.

4. Respecting the Free Flow of Information
Governments should permit the transfer of data and should not inhibit access by companies or individuals to lawfully available information that is stored outside of the country. Governments should not require service providers to locate infrastructure within a country's borders or operate locally.

5. Avoiding Conflicts Among Governments
In order to avoid conflicting laws, there should be a robust, principled, and transparent framework to govern lawful requests for data across jurisdictions, such as

improved mutual legal assistance treaty—or "MLAT"—processes. ... [I]t is incumbent upon governments to work together to resolve the conflict. ("Reform Government Surveillance" 2015)

It is interesting to notice how these principles model the FIPs—practices that the commercial enterprises themselves generally do not follow. The US law enforcement and intelligence communities insist "it's important that we do not let these technological innovations undermine our ability to protect the community from significant national security and public safety challenges" (Deputy Attorney General Sally Q. Yates in Senate testimony, quoted in Apuzzo, Sanger, and Schmidt 2015). At the same time, US tech companies wanting to enact (or enacting) stronger encryption are supported by, among others, the United Nations:

> Encryption and anonymity, and the security concepts behind them, provide the privacy and security necessary for the exercise of the right to freedom of opinion and expression in the digital age. Such security may be essential for the exercise of other rights, including economic rights, privacy, due process, freedom of peaceful assembly and association, and the right to life and bodily integrity. (Kravets, "UN Says" 2015)

Even some of the government's top intelligence people doubt the value of enabling government to defeat the encryption that is used commercially. For example, Michael Chertoff, former head of the Department of Homeland Security who was also a federal prosecutor, argues against cryptographic backdoors that could be provided to the government upon request, saying that "I think that it's a mistake to require companies that are making hardware and software to build a duplicate key or a backdoor even if you hedge it with the notion that there's going to be a court order" (Farivar, "Even Former Heads" 2015). In response to a dust-up in the cryptography community over a cryptographic trapdoor identified during the development of an encryption algorithm, the former director of NSA Research Michael Wertheimer noted in February 2015 that the "NSA should have ceased supporting the dual EC_DRBG algorithm [the trapdoor] immediately after security researchers discovered the potential for a trapdoor." He described the NSA's failure to cease supporting the algorithm as "regrettable," adding that "[t]he costs to the Defense Department to deploy a new algorithm were not an adequate reason to sustain our support for a questionable algorithm" (Wertheimer 2015: 165). The NSA failed to oppose the algorithm from 2007, when the weakness was first identified by Microsoft cryptographers, through 2013.

> [T]he consensus that strong encryption is good for security, liberty, and economic growth has come under threat in recent years. The June 2013 revelations about the U.S. National Security Agency's pervasive surveillance programs—not to mention the NSA's direct attempts to thwart Internet security to facilitate its own spying—dramatically shifted the national conversation, highlighting the vulnerabilities in many of the tools and networks on which we now rely for both everyday and sensitive communications. (Kehl, Wilson, and Bankston 2015: 21)

Cryptography's relevance and importance was highlighted by the awarding of the 2016 Turing Prize to John McCarthy and Whitfield Diffie, two "fathers" of American commercial cryptography (Markoff 2016). Disagreements between the government and tech companies leading to court battles—including those involving increased encryption on commercial products such as the iPhone (Nakashima 2016)—will play out for years (decades?) to come. While government representatives believe that "the technologists just hadn't tried hard enough to find a way to give government access to encrypted systems" (Abelson et al. 2015), privacy advocates note the challenges presented in the current environment and lament the government's demands that it be granted the keys to the backdoor, weakening commercial encryption:

> The complexity of today's Internet environment, with millions of apps and globally connected services, means that new law enforcement requirements are likely to introduce unanticipated, hard to detect security flaws. Beyond these and other technical vulnerabilities, the prospect of globally deployed exceptional access systems raises difficult problems about how such an environment would be governed and how to ensure that such systems would respect human rights and the rule of law. (Abelson et al. 2015)

Apple's response to increasing government demands that the company provide user data, whether those requests are from law enforcement, from the intelligence agencies, or mandated by court orders, represents a leading edge within the commercial tech community. However, Apple's response is not a radical departure from the themes underscored in the history of the 'Crypto Wars' between the government and commercial cryptographers. Apple's 'solution' to the problem is to make end-to-end encryption available to its customers. Under the protocols of Apple's latest operating systems, consumers who want to lock their systems, and who do not back up their systems to cloud-based services, prevent Apple and others—including law enforcement and intelligence operatives—from accessing data on their devices. Other high-tech manufacturers are moving in this direction, and the law enforcement and intelligence communities strongly condemn the practice. The government sees unbreakable, end-to-end encryption as a security threat that might be leveraged by criminals, terrorists, or other enemies. Yet, in the face of ubiquitous government requests and a degree of overreach (not to mention aspects related to the Snowden revelations), the commercial tech community moved to support Apple's stance:

> Google, Amazon, Facebook, Microsoft and a parade of other technology companies filed a barrage of court briefs on Thursday, aiming to puncture the United States government's legal arguments against Apple in a case that will test the limits of the authorities' access to personal data. In all, around 40 companies and organizations, along with several dozen individuals, submitted more than a dozen briefs this week … challenging every legal facet of the government's case, like its free speech implications, the importance of encryption and concerns about government overreach. … [T]he tech companies said they shared the public's outrage over the "heinous act of terrorism" in San Bernardino, but said they were united in the view that the government's case exceeded the boundaries of existing law and would hurt Americans' security. "We're facing a very big question as a country, industry and a world about what privacy will look like in the digital era," said Aaron Levie, the chief executive of the data storage company Box, which signed on to the brief with Amazon, Google and others. "There is a global impact for these tech companies if we don't land on the right side of having a strong framework for how companies deal with security and these kinds of requests in the digital age." Apart from the tech companies, seven prominent security experts and 32 law professors signed on to joint briefs on Thursday. Several industry trade organizations and digital rights groups submitted their own filings this week. Some echoed Apple's slippery slope argument that opening up one iPhone would lead to a domino effect from governments worldwide. (Wingfield and Bender 2016)

Although disagreements between the US government and high-tech industries about encryption standards have long-standing historical roots, the realities of day-to-day business and the global environment have reflected ubiquitous cooperation. Even in the face of disagreements about protecting US citizens' data, businesses with overseas markets must license their exports through the Commerce Department's export control list. While philosophical and political arguments about cryptography and encryption take place intermittently, the outcomes of negotiations and legislation are continuously enacted via export controls.

Encryption and Export Controls

Especially considering rapidly expanding global technology markets, US companies rely heavily on exports to produce broader sales and profits than domestic customers alone provide. Exportation of hardware/software containing and/or using encryption must pass muster under regulations of the exportation of technology products. Exports must be cleared through the export control list via a general license. "[C]ertain technologies are subject to export controls and companies that export them without a license can face fines and other penalties. A crucial determination for an exporter is whether its products use encryption technology since certain technologies for encryption cannot be transferred outside the United States" (Crane 2001). The export control list has long served as leverage for the US government, and especially the intelligence community, against commercial entities that would exceed and bypass the government's technical cryptographic capabilities. Long-standing agreements found the government demanding that only weak encryption may be exported even when stronger encryption was used in domestic products and services. Both the increase in global markets and the emergence of Internet-based commerce put extensive pressures on control procedures.

High-tech products and services are leading items on the control list. The Commerce Control List (CCL) "is divided into ten broad categories, and each category is further subdivided into five product groups": categories include (a) electronics, (b) computers, (c) telecommunications, and (d) information security; product groups include (a) systems, equipment, and components, (b) software, and (c) technology ("Commerce Control List" n.d.).

> The U.S. government regulates the transfer of information, commodities, technology, and software considered to be strategically important to the U.S. in the interest of national security, economic and/or foreign policy concerns. There is a complicated network of federal agencies and inter-related regulations that govern exports collectively referred to as "Export Controls." In brief, Export Controls regulate the shipment or transfer, by whatever means, of controlled items, software, technology, or services out of [the] U.S. ("Export Control" n.d.)

The development of various levels of export controls took place over an extended period of time, from the very onset of computational, networked new media. Early contestations, accommodations, and agreements moved toward settling some of the questions at hand. As explained by Kehl, Wilson, and Bankston:

> While the domestic fight over key escrow wore on throughout the mid-1990s, another related battle was brewing on the international front over U.S. export controls and encryption technology. The question at the center of that debate was whether American technologies containing strong encryption should be made

available overseas—which would in turn have a significant effect on the domestic availability and use of encryption tools. Until 1996, cryptographic tools were classified as munitions in the United States, with strict limits on the type of encryption that could be exported and the maximum cryptographic key length. Despite growing opposition to these restrictions, the U.S. government had a strong incentive to maintain encryption export controls as a means to delay the spread and adoption of strong encryption technology abroad. The practical result of the policy was that many companies exported weaker versions of their encrypted products, or were kept out of foreign markets altogether. By the mid-1990s, experts projected billions of dollars in potential losses as a result of these policies. Coupled with growing evidence that foreign-made encryption was readily available around the world, the rationale behind maintaining these controls became increasingly tenuous.

Many of the same organizations and individuals that rallied against the Clipper Chip came together to mobilize against encryption export controls, arguing that they undermined U.S. economic competitiveness and individual privacy, with little evidence that they were actually achieving their stated goals. From 1996 to 1999, the Clinton Administration gradually liberalized encryption export controls, beginning with the 1996 Executive Order that moved most commercial encryption tools from the U.S. Munitions List to the Commerce Control List. The next step involved relaxing limits on the strength of encryption keys. Although these concessions were originally used as a bargaining chip in the commercial key escrow debate—companies would be allowed to export higher strength encryption if they agreed to retain the keys—those requirements were eventually abandoned after pressure from industry and public interest groups. In September 1999, the White House announced a sweeping policy change that removed virtually all restrictions on the export of retail encryption products, regardless of key length. As journalist Steven Levy put it succinctly: "It was official: public crypto was our friend." In the decades since the resolution of the Crypto Wars, many of the predictions about how strong encryption would benefit the economy, strengthen Internet security, and protect civil liberties have been borne out. In particular, the widespread avail[ability] of robust encryption laid the groundwork for the emergence of a vibrant marketplace of new Internet services based on secure digital communications and the widespread migration of sensitive communications online. (Kehl, Wilson, and Bankston 2015: 3–4)

Non-military cryptography exports from the United States are controlled by the Department of Commerce's Bureau of Industry and Security (BIS). Registration is required for mass-market products using strong encryption. Some other products—for example, open-source cryptographic software—also have to be registered (but in some cases, without review). US non-military exports are controlled by Export Administration Regulations. Although the various categories and requirements are exceedingly complex (thus making blanket determinations risky) protecting media content via standard digital rights management is excluded from consideration under the control lists ("Identifying Encryption Items" n.d.).

Export controls on encryption technologies present a major conundrum to all parties interested in privacy, especially in consideration of protecting data from multiple uses. Privacy advocates in the United States are generally supportive of the efforts of overseas entities (be they courts, governments, or individuals) to protect individual privacy. Generally speaking, strong encryption enables control of data and therefore undergirds personal privacy. Resistance by US government to the exportation of strong encryption undermines this goal. Likewise, US intelligence work raises major concerns regarding the multiple uses made of personal data. Strong encryption would go a long way toward limiting the ability of intelligence services, belonging to any country, to extract meaningful content from intercepted transmissions.

One can expect that, at some point in the near future, contestation over strong encryption will shift from skirmishes over individual devices or incidents to once again focus on export controls. As long as the government can limit the exportation of strong encryption, US companies will be stuck in an unenviable situation: They will be unable to prevent US spy and law enforcement from using consumers' personal data and they will be equally unable to provide overseas markets with the level of encryption wanted by foreign partners, thereby weakening the abilities of our tech producers to compete in the global marketplace. Something will have to give.

The Snowden Revelations

Verdicts as to whether Edward Snowden is a patriot or traitor, a whistleblower or criminal, to be praised or reviled, will depend on future actions taken by politicians and courts. This volume contends that many of the Snowden revelations validated facts that had appeared in the press before or after the release of the classified materials provided by Snowden. There appears to be little doubt that Snowden broke laws by releasing classified materials. Whether the release of those materials was justifiable under whistleblower protections remains to be seen.

The materials in the Snowden revelations were so spectacular that they outlived the typically short news cycle. Significant numbers of news reports appeared throughout the first year after the stories broke. The issues still command attention to this day. Much about the situations that Snowden either revealed or verified remain problematic and, to some degree, unchanged. For example, materials presented below describe the ways that the courts and Congress have enabled the NSA to continue most of the data collection programs that were made public. Some observers hoped that making public the facts about those programs would lead to their curtailment. Likewise, especially in the area of intelligence gathering but also with regard to law enforcement, governmental entities continue to use the data streams produced by commercial activities, such as the use of mobile phones and the World Wide Web. Sometimes governmental data gathering is the result of willing cooperation between commercial entities and the government. At other times, courts force that cooperation.

However, the public awareness that resulted from the revelations placed the commercial outfits in a new position. Cooperating with the government in secrecy is one thing; being seen as the hidden arm of the government is quite another. The most defensible interpretation of the outcomes of the Snowden revelations appears to be that taking data collected for one purpose and putting them to other uses is a practice now open for public interest and debate. Faced with this public conversation, some of the commercial entities appear less inclined to continue cooperating with the government in the same ways as they did before Snowden's revelations.

By 2015, the courts and Congress had reinstated most of the data collection programs run by the NSA, including those that collect domestic information (Gershman, "Read the Ruling" 2015). In June of that year, Congress passed the USA Freedom Act, which reinstated, with some new practices, provisions of the Patriot Act that had expired. The biggest change requires that telecommunication companies store the data that the NSA scours. However, the essentials of the surveillance and data collection programs were changed little by the act (Farivar, "Even Former NSA" 2015).

Regardless of attempted constraints by courts and legislation, the NSA finds ways "to create a functional equivalent. The shift has permitted the agency to continue analyzing social links revealed by Americans' email patterns, but without collecting the data in bulk from American telecommunications companies—and with less oversight by the Foreign Intelligence Surveillance Court" (Savage 2015). While seeking increased abilities to mine the social media data of foreigners visiting or applying for entry into the United States, the intelligence community, at the behest of Congress, continues to develop collection and analytic tools that can be used to "fully automate the process of going through a huge amount of messages and other data" (Nixon 2016).

Security expert David Heyman warns that "You have to be careful how you design the proposal to screen people. ... Artificial intelligence and algorithms have a poor ability to discern sarcasm or parody" (quoted in Nixon 2016).

While little has changed on the government side with regard to making use of American citizens' private data in ways other than those that were intended when the data were provided, significant changes have occurred on the commercial side. Edward Snowden notes, "The biggest change has been in awareness. Before 2013, if you said the NSA was making records of everybody's phone calls and the GCHQ was monitoring lawyers and journalists, people raised eyebrows and called you a conspiracy theorist" ("Edward Snowden" 2015). The encryption controversy raised by Apple's refusal to cooperate with the FBI illustrates the growing awareness among technology firms of the precariousness of their situation. Now that there is more public awareness, continued cooperation with the government's data collection programs suggests that commercial entities either cannot or will not protect the privacy of consumer data. This compromises corporate assurances that they can and will protect privacy. Their failure to push back against government demands might lead to a lack of trust on the part of consumers.

The Apple-iPhone dust-up pitted big tech against big government. Apple, according to the Obama administration, sided with "its business model and public brand marketing strategy" ahead of public safety (Ackerman and Yadron 2016). That's not it, says Apple CEO Tim Cook. He says his company is "a staunch advocate for our customers' privacy and personal safety" (quoted in Rall 2016). On this aspect, Rall notes:

> Rather than face Uncle Sam alone, Apple's defiance is being backed by Facebook, Google, Microsoft, Twitter and Yahoo—companies who suffered disastrous blows to their reputations, and billions of dollars in lost business, after NSA whistleblower Edward Snowden revealed that they spent years voluntarily turning over their customers' data to the spy agency in its drive to "hoover up" every email, phone call, text message and video communication on the planet, including those of Americans. Most Americans tell pollsters Apple should play ball with the FBI. But Apple and its Silicon Valley allies aren't banking on the popular vote. Their biggest customers are disproportionately well-off and liberal—and they don't want government spooks looking at their personal or business information. ... The NSA, specifically chartered to intercept signals intelligence that originates overseas—that is specifically prohibited from gathering data that is sent from one American to another American—continues to do so, probably at an even greater degree of efficiency than the period between 2009 and 2013, the era documented by the Snowden revelations leaked to the news media. Ignoring the anger of the American people, Congress did nothing to rein in the NSA. (Rall 2016)

Former CIA agent and fiction author Barry Eisler notes, "So much of Snowden's revelations were about this very thing. And the fact that the public knows about corporate cooperation with the government now is in part, I think, what has emboldened Apple to push back. ... If we didn't know about these things, I would expect that Apple would be quietly cooperating. There would be no cost to their doing so" ("Former CIA" 2016).

Data Collection and Insecurity: The Government Sharing with Commercial Entities

Records collected by local, state, and federal governments ('the government' or 'public entities') follow citizens from birth to death. These records include vital records (e.g., birth, marriage, divorce, death), work history/professional records (e.g., licensing, workers compensation), home and/or property records (e.g., tax assessments, transaction prices), and law enforcement and court records (both civil and criminal). Many of these records contain information that is linked to, or contained in, other records, thereby providing additional data and less control than for material in a single collection. For example, records accumulated in a civil court case will refer to information that was introduced and might include medical information, lifestyle information, travel habits, and so on. Home and property records can include information that specifies elements of the dwelling or the property such as security systems, smart appliances, swimming pools, and health-related spas. Work history records include date ranges for employment, workplace locations, and other specifications that can increase the accuracy of data mining.

Federal law, federal agency policies, and the Freedom of Information Act (FOIA) manage access to records collected by the federal government. Generally, individuals are able to access their own records by following agency procedures. Other parties sometimes access federal documents via FOIA requests. State governments have passed open records laws that resemble the federal FOIA statute. However, "[s]tate FOIAs generally do not permit any discrimination among requesters. In a number of cases, officials wanting to restrict access to people requesting records for commercial use had no statutory authority to do so" (Solove 2004: 151). Most public entities have established procedures for ordering, and in some cases paying for, the administrative work required in meeting requests for records. Many states have passed legislation specifying costs for records recovery in the face of numerous requests for information from the commercial sector. State governments want to avoid incurring excessive costs for doing the business of government at the behest of commercial entities wanting to leverage public information. Fulfilling data requests therefore provides an additional revenue stream to governmental entities; financial gain sometimes encourages shoddy decision making about records releases. For example, many states continue to earn millions of dollars annually by selling information contained in driver's license records ("Jacobs Calls" 2015).

Numerous public entities provide access to public data. For instance, the Gallagher Law Library, University of Washington School of Law, offers access to the following prominent sources:

- campaign finance disclosure data search from the Federal Elections Commission: searchable
- charities and nonprofits (exempt organizations) from the Internal Revenue Service: searchable
- corporation filings from the Securities and Exchange Commission (EDGAR): searchable
- hazardous waste sites from the Environmental Protection Agency (Superfund): searchable
- HUD/FHA refunds from the U.S. Department of Housing and Urban Development
- inmate locator from the Federal Bureau of Prisons: searchable, from 1982
- patents from the Patent and Trademark Office: searchable
- sex offender registry from the Justice Department, with all 50 states, U.S. Territories, the District of Columbia, and participating tribes represented as of January 2012
- toxics release inventory (release of toxic chemicals into the environment) from the Environmental Protection Agency: searchable
- trademarks from the Patent and Trademark Office: searchable
- trademark dispute documents from the Patent and Trademark Office
- Investigative Resource Center
- Public Record Locator (*US Public Record* 2014)

Gaining access to, and utilizing, public records is not only part of the professionalized data marketplace, it is also a highly sought-after skill among businesspeople who do not want to pay for data marketplace services. Curtis Frye's (2015) online course "Up and Running with Public Data Sets," for one, teaches students "how to find free, public sources of data on a variety of business, education, and health issues and download the data for [their] own analysis." Specifically, Frye introduces students to "resources from the US government (from census to trademark data), international agencies such as the World Bank and United Nations, search engines, web services, and even language resources like the Ngram Viewer for Google Books."

One of the shining promises of the early Internet was that enhanced database connectivity would increase citizens' access to public records, thereby opening government to public scrutiny. There is no doubt that the contemporary new media environment offers increased access to a wider amount of public information to a larger number of people than ever before—and that, to an extent, 'sunshine is a great disinfectant.' However, numerous kinks exist in the public information value chain.

> Marketers stock their databases with public record information, and the uses to which these databases are put are manifold and potentially limitless. The personal information in public records is often supplied involuntarily and typically for a purpose linked to the reason why particular records are kept. The problem is that, often without the individual's knowledge or consent, the information is then used for a host of different purposes by both the government and businesses. ... The problem is not necessarily the disclosure of secrets or the injury of reputations, but is one created by increased access and aggregation of data. ... Do we want to live in a Kafkaesque world where dossiers about individuals circulate in an elaborate underworld of public- and private-sector bureaucracies without the individual having notice, knowledge, or the ability to monitor or control the ways the information is used? (Solove 2004: 149)

In many instances it is difficult to limit access to private information, from public databases, to the individuals who produced the data. Sometimes the data can be used against the very people who produce it. Two examples illustrate this dilemma. First, lawyers have discovered that extensive materials are available online and may be accessed during case discovery procedures. Obviously, there are two sides in each case; one side is happy that their lawyer can access a lot of the data from people on the other side, and the reverse is also true. The second instance illustrates ways that even everyday private citizens can leverage access to public data in order to harm the people who provided the data in first place.

Writing for lawyers on the concept of "e-discovery," Helen W. Gunnarsson (2010) notes, "An amazing amount of information about parties, witnesses, clients, businesses and more resides on the Internet. And much of the best stuff doesn't cost a dime." She quotes litigator Todd Flaming, who differentiates between what he calls Discovery 1.0 and Discovery 2.0: "You've now provided me with free discovery. Not e-discovery, *free* discovery. ... All the people you are dealing with have posted something on the Internet that is helpful to you, if not because it's relevant evidence, at least for giving you helpful background and getting you started." Another Chicago-based litigator, Kent Sezer, demonstrated how public documents such as home purchase data, tax records, mortgage documents, and bank records can help establish patterns of behavior that can be used to undermine litigants' claims of poverty or inability to marshal financial resources (in Gunnarsson 2010). In effect, information about people that exists in the public domain can be, and sometimes is, used against those individuals.

A second instance illustrates a very troublesome fact about publicly available information: Not only can it be used against the individual who provided it, but it can be used maliciously as a way to influence behavior or to punish parties for their behaviors or activities. Further, material that qualifies as public data sometimes doesn't appear to be public data at first glance. For example, individuals who work at institutions affiliated with or sponsored by local state and federal governments produce enormous amount of data, much of which can be classified as public data. A specific example illustrates: Electronic mail messages produced by faculty members at state universities count as public documents. The case of William Cronon, a University of Wisconsin professor, demonstrates how easy it is to turn public documents into blunt instruments against the people who produced the data:

> Cronon has been a vocal critic of Gov. Scott Walker's crusade against collective bargaining. … Stephan Thompson of the Wisconsin Republican Party filed an open records request for the professor's personal e-mails—including any memorandum that contains the words "Republican," "Scott Walker" and "union," among others. … University of Wisconsin-Madison Chancellor Biddy Martin released a statement in response to the controversy, saying that the university would comply with the open records request as required by law. (Finnegan 2011)

The enormous amount of publicly available information that can be acquired and potentially used against citizens raises the degree of concern one might have over the need to provide the documents publicly in the first place. While it is the case that an individual should be able to recover the data within their records, it is not always equally the case that others should be able to do so. Complex circumstances are in play, of course.

For example, it is socially useful for people wanting to buy homes to be able to access public real estate documents. It is exceedingly useful for companies doing business with citizens to be able to find customers who have changed their domicile and address. In each case—property transaction prices and tax records in the first instance and change-of-address records held by the post office in the second instance—governmental entities collect information and make it available to the public (including the people who generated the data in the first place). However, those same records are exceedingly important to companies within the data marketplace. Being able to target consumers requires finding them; tax records and change-of-address forms are key indicators that data aggregators purchase from the government. Companies within the data marketplace purchase information from data wholesalers; those outfits—for example, AccuData and CoreLogic—purchase data sets directly from government agencies (Tanner 2014: 62–63). Data wholesalers purchase public record data from county offices, state offices, and the federal government. For example, "Inflection spends between $3 million and $5 million a year to buy and rent [the] personal data … [f]rom outfits such as Axiom, Epsilon, … Experian, TransUnion, and Equifax" (Tanner 2014: 63). The United States Postal Service is also an important player in the data broker marketplace.

> Whenever a person changes an address or forwards mail, he or she signs a form whose fine print authorizes the USPS to share the information with companies "already in possession of your name and old mailing address." Such information allows data brokers and companies to update older lists continuously. (Tanner 2014: 65)

Material from public sources can include those US Postal Service records; state and county property records; corporate filings and Securities and Exchange Commission data; information given to public schools, colleges, and universities that is not protected under specific privacy laws; county, state, and federal court office records; professional licenses and accreditation; and in some states, information from state license plate readers, toll booths, and/or motor vehicle department records.

In many cases the public good is served by open accessibility to public records. Records provided to governmental entities should be open to the citizens who generated the data; governmental entities that process the data are paid for through public funds derived from taxpaying citizens. The ability of citizens to access their records comports with the second FIP. The ready entry of those data into the commercial marketplace presents conundrums for the public sphere and challenges to government officials and privacy advocates alike:

The threat to privacy is not in isolated pieces of information, but in increased access and aggregation, the construction of digital dossiers and the uses to which they are put. States must begin to rethink their public record regimes, and the federal government should step in to serve as the most efficient mechanism to achieve this goal. It is time for the public records laws of this country to mature to meet the problems of the Information Age. (Solove 2004: 161)

Crossing Government and Commercial Streams within the Data Marketplace

Daniel Solove focused an insightful and prescient chapter on the government's use of data collected in the commercial marketplace in his 2004 book, *The Digital Person*. Drawing on material and arguments presented throughout the book, Solove made clear, 12 long years ago, that government access to the breadth and depth of private information available on the Internet would deeply compromise democracy and the rule of constitutional law in the United States. Solove detailed the many ways that data collected for one purpose (by both commercial and governmental entities) is accessed and used by the government for other uses.

Solove presented the numerous collections of records, commercial and governmental, that are of interest to law enforcement and intelligence communities. Records held by third parties are particularly interesting because they are less protected by law than is information taken directly from an individual. Solove reminded readers about the staggering breadth of the materials within one's commercially derived digital dossier. He presented what was then a recent controversy, over the Department of Defense's Total Information Awareness (TIA) program. After TIA was exposed by the press, Congress hastily did away with it. However, TIA was very similar to the extensive data collection and analysis operations that later became known as "The Program"—operations that stand at the base of the Snowden revelations and continue through the present. Solove also identified a number of private entities within the data marketplace, including ChoicePoint and SeisInt, that had established contractor relations with governmental entities in order to provide access to third-party data.

Solove discussed intricacies in the legal relationships that developed in the early days after 9/11, focusing on the degree to which new laws, practices, and court orders sometimes required companies to cooperate with the government by providing third-party data. He reviewed some of the accommodations made by new media businesses within their ToS and/or privacy policies. Solove argued that "businesses and government have become allies":

> When their interests diverge, the law forces cooperation. The government can increasingly amass gigantic dossiers on millions of individuals, conduct sweeping investigations, and search for vast quantities of information from a wide range of sources, without any probable cause or particularized suspicion. Information is easier to obtain, and it is becoming more centralized. The government is increasingly gaining access to the information in our digital dossiers. ... Thus, we are increasingly seeing collusion, partly voluntary, partly coerced, between the private sector and the government. While public attention has focused on the Total Information Awareness project, the very same goals and techniques of the program continue to be carried out less systematically by various government agencies and law enforcement officials. We are already closer to Total Information Awareness than we might think. (Solove 2004: 174–175)

Without exaggerating, one could claim that most of the worst case scenarios presented in Solove's 2004 treatise have come to pass. With increasing frequency, the US government taps into commercial entities' databases and networks for access to information about American citizens.

Reports published in 2013 indicate a strong push, on the part of US military and intelligence entities, toward collecting and analyzing private data held by third parties. Bloomberg reported that, "[a]rmed with billions of tweets, Google searches, Facebook posts, and other publicly available social-media and online data, the Office of the Director of National Intelligence is sponsoring research projects involving 14 universities in the United States, Europe, and Israel with the goal of using advanced analytics to predict significant societal events" (Warner 2013).

By May 2013, reports connected to the information provided by Snowden began to appear, although journalists Glenn Greenwald and Laura Poitras did not arrive in Hong Kong to meet with Snowden until June 2 of that year. James Risen and Laura Poitras published an article on May 31, in the *New York Times*, indicating that the NSA collects massive amounts of facial imagery from the Web. By September, key reporters in the Snowden affair had published numerous articles detailing NSA data collection and analysis:

> Since 2010, the National Security Agency has been exploiting its huge collections of data to create sophisticated graphs of some Americans' social connections that can identify their associates, their locations at certain times, their traveling companions and other personal information, according to newly disclosed documents and interviews with officials. The spy agency began allowing the analysis of phone call and

e-mail logs in November 2010 to examine Americans' networks of associations for foreign intelligence purposes after NSA officials lifted restrictions on the practice, according to documents provided by Edward J. Snowden, the former NSA contractor. (Risen and Poitras, "NSA Gathers" 2013)

Throughout 2013, numerous articles and analyses added credibility to claims substantiated by the documents revealed by Snowden. Bamford (2012), Goodin ("Contrary" 2013), McChesney (2013), and Warner (2013), among others, published materials drawing attention to government use and misuse of private data. That year also witnessed the passage of laws in a number of state legislatures protecting the privacy of workers' social media accounts in workplaces (Milligan, "California" 2013, "Michigan" 2013; Rubin and Stait 2013; Schaefers, "Nevada and Colorado" 2013, "Washington State" 2013).

By 2014, research by LexisNexis indicated that "[e]ight out of every 10 law enforcement professionals (81%) actively use social media as a tool in investigations. … The frequency of social media use by law enforcement, while already high, is projected to rise even further in the coming years." Unfortunately, the same research indicated that, "[d]espite widespread use of social media for investigations, over half (52%) of the agencies surveyed do not have a formal process in place regarding the use of social media for investigations. … [F]ew agencies have adopted formal training, policies or have dedicated staff in place, resulting in barriers to consistent and broad application throughout all of law enforcement" ("Social Media Use" 2014: 8). In the face of these weaknesses in training, policies, and staffing, law enforcement officials continued to call for increases in the use of social media–based analytics:

> The increase in use of interactive social media in the last 5 years has changed how people live their lives. Individuals use smart phones to post photos, identify their location, and advise what they are doing. If law enforcement agencies fully engage the public by using social media, the department and the community benefit by increasing collaboration and enhancing investigative capabilities. With well-planned implementation, the use of social media can impact community issues, and police departments and citizens can work together to solve crimes. (Henton 2013)

Contrary to the hopes of privacy advocates in the wake of the Snowden revelations, mining of commercial, third-party data by law enforcement and the intelligence community increased after 2013. In an article published in 2014, the *Wall Street Journal* reported:

> For the past 18 months, the U.S. has invested heavily in ways to collect and examine social-media postings on Facebook, Twitter and overseas regional networks as a source

of overseas intelligence, according to Gen. [Michael] Flynn and other officials. ... [G]overnment computers can aggregate material from multiple social-media networks and scan massive amounts of information already publicly available to any computer user for trends and links. ... The new generation of tools allow U.S. spy agencies to look for trends and warning signs of potential national security crises, Gen. Flynn said. They allow intelligence agencies to zero in on social-media postings during a set period from a defined geographic area spanning a few square miles, an entire country, or a continent. ... Analysts also can take a Twitter message or other posting and uncover connections between the author and others in the social-media universe; once they find a target's contacts, analysts can determine the physical location of the contacts—in essence applying the social media network to a real-world map. ... The DIA also has developed a tool that scans social media for individual faces—sifting millions of postings for images of a single person. (Barnes 2015)

In October 2014, an *NBC News* headline noted that the "FBI Trolls Social Media for Would-Be Jihadis" (Windrem and Brunker 2014). Some privacy advocates noted (with reproach) that this program uses tactics that misrepresent the identities of government participants, perhaps enticing targets to do things that they might not otherwise have done.

Although the crossing of government and commercial data streams causes a significant number of problems, two particularly important—fundamental and broad—aspects must be noted. First, contemporary data mining may well further damage already disadvantaged populations. Second, data mining exacts a particularly unfair toll on all consumers within capitalist economies.

Civil rights organizations are challenging the extensive uses of big data analysis by government and commercial entities alike, on the premise that the practices can bear special harms for disadvantaged populations. From one point of view, data mining produces a sort of high-tech profiling; there are risks that the marketplace will tailor itself to only those who can afford products and services and will specifically avoid those who cannot, including minorities and low-income citizens. Turow notes that

> marketers are increasingly using databases to determine whether to consider particular Americans to be targets or waste. Those considered waste are ignored or shunted to other products the marketers deem more relevant to their tastes or income. ... The quiet use of databases that combine demographic information about individuals with widely circulated conclusions about what their clicks say about them present to advertisers personalized views of hundreds of millions of Americans every day without their knowledge. ... The unrequested nature of the new media buying routines in the directions these activities are taking suggest that narrowed options and social discrimination might be better terms to describe what media buyers are actually casting. (Turow 2011: 88–89)

Wade Henderson, chief executive of the Leadership Conference on Civil and Human Rights, claims that "[b]ig data has supercharged the potential for discrimination by corporations and the government in ways that victims don't even see. ... This threatens to undermine the core civil rights protections guaranteed by the law in ways that were unimaginable even in the most recent past" (quoted in Fung, "Why Civil Rights" 2014). Data collection and the data marketplace potentially reach all Americans, but minorities and low-income citizens face special risks in this regard; big data and civil rights are about targeting and separating one type of individual from another. Matthew Crain reminds us that "sorting is as much a practice of exclusion as inclusion, and the broader and more integrated the sorting becomes, the greater the potential for discriminatory practices"; he points out that a 2012 *Wall Street Journal* investigation "showed that internet retailers routinely engage in price discrimination based on information obtained from consumer surveillance" (Crain 2013: 187–188).

Ryan Calo (2014) identified one of the most insidious and widespread negative aspects of putting data to purposes other than those for which it was collected: digital market manipulation. While it is not the case that every market within a capitalistic system must be fair (there is an expectation that some markets will be more fair than others and that some markets will treat some consumers in special ways), there is an expectation that the market in general will not be unfair to virtually everyone. Calo describes both objective and subjective harms to consumers from digital data market manipulation. First, subjectively,

> the consumer has a vague sense that information is being collected and used to her disadvantage, but never truly knows how or when. In the digital market manipulation context, the consumer does not know whether the price she is being charged is the same as the one charged to someone else, or whether she would have saved money by using a different browser or purchasing the item on a different day. The consumer does not know whether updating his social network profile to reflect the death of a parent will later result in advertisements with a heart-wrenching father and son theme. She does not know whether the subtle difference in website layout represents a "morph" to her cognitive style aimed at upping her instinct to purchase or is just a figment of her imagination. (Calo 2014: 1029)

This experience presents the consumer with a Kafkaesque dilemma brought about when data they have provided to specific entities is circulated among and misused by a broad range of unknown entities.

Objectively, digital market manipulation finds consumers at risk of paying more for products and services rather than less. This practice seems contrary to everyday experience, as many online shoppers find 'deals' galore and would credit the online environment with enabling them to comparison shop for the lowest prices. In the limited context of their online shopping experience, these consumers may save money in the short term, thereby validating their perception. However, in the broader marketplace, consumers may well give back those gains because digital market manipulation uses personal information to "extract as much rent as possible from the consumer" (Calo 2014: 1001). In the long run, consumers lose because "the consumer is shedding information that, without her knowledge or against her wishes, will be used to charge her as much as possible, to sell her a product or service she does not need or needs less of, or to convince her in a way that she would find objectionable were she aware of the practice" (Calo 2014: 1030).

In short, extensive exchange of private data within the data marketplace—exchange that almost always includes data sourced from government and commercial entities—has the strong potential to damage the very providers of that information in violation of the third FIP.

We Claim to Know Better but Do Not Act Like It

To conclude this chapter, then, we return to a key feature of the third FIP: Regardless of what governmental and commercial entities do with personal data, citizens who provide the data bear responsibilities for its existence in the marketplace. While citizens deserve some degree of protection from data marketplace entities, outfits that can and do misuse private information, the first line of protection lies with citizens, who sometimes are careless with what might otherwise be private data.

One can argue that citizens of the United States make themselves particularly vulnerable to mass data collection both by governments and by commercial entities. Security assessment specialist Marc Quibell (2013) notes that "people still freely give out information, more information than ever before, more information than the NSA could ever get and at real-time speeds, to entities who deal in collecting, trading and selling personal information. Entities who also tend to lose that information to cyber criminals." Quibell wonders why "people have become more comfortable with companies like Facebook or LinkedIn 'spying' on them, and consider the NSA 'spying' an affront to civil liberties." Referring to the settings that are available in popular smartphone apps, he recounts the permissions for data collection that we give to app developers and providers. Specifically, you likely allow social media companies "to read data about your contacts stored on your phone, including the frequency with which you've called, emailed, or communicated in other ways to specific individuals"; "to modify and save your call log (not to mention just read all your call logs)"; "to share and save your calendar data"; "to directly call phone numbers" (which costs you money); to access recording devices; and "to know your location" (Quibell 2013).

Fundamental tensions are at play here. Americans have, for decades and relatively uniformly across age groups, reported in surveys that they value privacy. However, their behavior speaks volumes to the contrary, as they routinely provide enormous amounts of information to the data marketplace, particularly through social media. One tension, then, is between what people say and what they actually do. A second tension is between what citizens feel they should be able to do—what they want to be able to do—and what they can in fact accomplish. Contemporary Americans have clearly shown that they want to participate in social media. They want to engage each other using digital communication technologies. They enjoy staying in contact with friends and family and in some cases they want to project themselves into broader social conversations with wider populations. They do not, however, participate in this sociality for the purpose of providing intelligence information to law enforcement or the government, nor are they eager to be tracked by, and be the victims of, target marketing from commercial operations.

On the one hand, it is rather easy to say that people should just stop using social media. After a brief period of time (about two years) during which I tested Facebook to learn about its functionalities as well as to help organize a high school reunion event, I canceled my Facebook account, primarily as a way to protect my privacy (Lamoureux 2014). From time to time I wish that I had not canceled the account. Often I find myself ignorant about events—past, present, or future—that my friends or family know about or have participated in. The situation seems even more dire to our children (now young adults and adults), who act as though they are at a total loss when beyond the reach of their Facebook feed. It seems unreasonable to expect that contemporary Americans will just turn social media off. And many would argue that they should not have to.

In a very short period of time, electronic media use and the misuse of personal data therein have become ubiquitous and deeply entwined. "Every keystroke, each mouse click, every touch of the screen, card swipe, Google search, Amazon purchase, Instagram, 'like,' tweet, scan—in short, everything we do in our new digital age can be recorded, stored, and monitored" (Harcourt 2015: 1). Furthermore,

> Our digital cravings are matched only by the drive and ambition of those who are watching. ... We face today, in advanced capitalist liberal democracies, a radically new form of power in a completely altered landscape of political and social possibilities. ... We are not so much being coerced, surveilled, or secured today as we are *exposing* or *exhibiting* ourselves knowingly, many of us willingly, with all our love, lust, passion, and politics, others anxiously, ambivalently, even perhaps despite themselves—but still knowingly exposing themselves. ... We live in what I would call the expository society. (Harcourt 2015: 14, 17, 18, 19)

The discussion in chapter 7 includes ways that entities such as social media firms might continue doing business while complying with the FIPs. Without changes in the business models and practices of social media firms, citizens are faced with the stark choice to lose control over their data by using social media or give up social media's perceived advantages in favor of privacy protection. Americans have shown no interest in the latter choice. So, while users who want to protect their privacy while also continuing to use social media *will* have to modify their behavior, sponsoring companies will also have to modify product functionalities. Limiting the use of personal data to the purpose(s) for which they are given is an important, and oft-violated, feature of protecting privacy in the age of new media.

· 6 ·
FIP 5

If You Don't Protect It, You Should Not Have It

> If by my life or death I can protect you, I will.
> —J.R.R. Tolkien, *The Fellowship of the Ring*

> Any organization creating, maintaining, using, or disseminating records of identifiable personal data must assure the reliability of the data for their intended use and must take reasonable precautions to prevent misuse of the data.
> —FIP 5, *Records, Computers, and the Rights of Citizens*

Violations of specifications in the second part of the fifth FIP—that holders of a private data point protect it to prevent its misuse—were illustrated in chapter 1. Further illustrations, from reports in 2016, will follow in the present chapter. First, however, this chapter explores a less-often examined aspect of data-privacy protocols, as suggested by the fifth FIP: Organizations must "assure the reliability of the data for their intended use." Without becoming strictly tied to statistical language, reliability of private data refers to the accuracy with which the data represent the subject(s) who provided the data.

We will, first, examine the ways that the data marketplace furthers the initial goal of the fifth FIP, improving the reliability of data. Analysis of large data sets provides insights that help solve problems and thereby giving benefits some of the people who contributed the information. Consumers often benefit from recommendations that are the result of analyzing data that they provided. In some ways, tracking and targeting consumers leads to more accurate inferences about them than was previously possible, thereby making advertisements and appeals directed at them more efficient and cost-effective for companies and more relevant and less objectionable to the people targeted. However, industry claims that the companies use consumer-generated data to improve usability and functionality are generally self-serving and self-fulfilling false prophecies. Second, continued shortcomings in marketing and advertising suggest that unreliability is too high to justify wide-ranging privacy violations. Third, specific misapplications of private data violate principles of the fifth FIP. Finally, this chapter adds details to the data breach material presented in chapter 1 and strongly suggests that the amount of private data put at risk by data breaches presents a strong argument against the viability of maintaining the status quo and toward supporting the fifth FIP.

Improved Data Reliability via the Marketplace

Privacy researchers have long known that citizens and consumers respond differently to the collection of their private information depending on whether they believe it is being put to relevant use. A 2007 study published in the journal *Marketing Letters* noted that "consumers are highly sensitive to whether we think our data is being used in ways relevant to the purpose at hand" (Scola 2014). Issues surrounding relevance underpin the theories of Nissenbaum (2010, 2015) and others about privacy and contextual integrity. When citizens feel that the data they have provided are put to the uses for which they were intended, or when they feel they have derived significant rewards from the uses to which the data are put, they often relax strict norms and rules that might otherwise define the collection as a privacy violation. The increased accuracy provided by big data analysis and by tracking and targeting sometimes authorizes acceptance of (and perhaps contentment with) marketplace uses of personal information.

Big Data Can Aid Problem Solving

The IoT and 'big data' analysis, in symbiosis, represent a broad range of potential advances and advantages. People experience big data analysis when they receive recommendations from online retail outlets as a result of using Internet search engines. These features will be discussed in the next section.

In these early days of developing the IoT, many of the IoT's advantages will accrue to corporate entities with the hope that subsequent advances will be passed along to consumers by way of improved products and services and/or lower prices. Even at this early stage, however, some direct-to-consumer advantages can be demonstrated.

In industrial applications (e.g., manufacturing), analytical systems can help operators make better decisions about logistics, plant configurations, component deliveries, and other key operational features, such that "analytics could make people more efficient by getting them where they need to be at the right time with the right tools" (Gallagher, "Machine Consciousness" 2015). Although some may worry about analytical systems putting humans out of work, there's a good chance that such systems can be used to augment and improve industrial performance in ways that lower costs to eventual end-product consumers. Further, analytical systems can help industrial entities make decisions that can decrease accidents and improve overall performance, thereby leading to safer work environments with higher productivity without displacing workers who are needed to implement, monitor, and perform maintenance on equipment and technological systems.

> For example, GE's Connected Experience Lab ... system pulled in not just information from the power grid, but weather sensor data and even information on historic and projected tree growth in areas around power lines to help predict in advance where there might be outages caused by wind or fallen branches. Similar analytic systems included geospatial data on railroad lines and track surveys, aiming to prioritize track maintenance to prevent derailments and other incidents. (Gallagher, "Machine Consciousness" 2015)

While it is the case that, currently, "many of the big data analytics and insights revolve just around operations ... [and it] is there that administrators revise, improve, and streamline workflows to reduce costs," such insights can also be used to deliver higher quality products and services on the way to improving customer satisfaction (Shah 2015).

The healthcare industries may well provide the best current instances of customer-specific outcomes from IoT applications. For example, "[t]here is a growing demand from consumers towards greater access to and portability of their health information" (Shah 2015). This demand can be met via cloud-based electronic health records. Numerous advances also exist in the area of wearable technologies and biosensors that monitor behaviors and physiological functions as well as treat specific conditions. Further, big data analytics at the point of care will "become an essential tool in the doctor's toolkit. Operationally, leveraging health insights and pivoting this knowledge into targeted patient outreach and access is what makes analytics so powerful" (Shah 2015).

Data analytics in health care illustrates many of the privacy promises, challenges, and conundrums faced by systems, providers, and consumers alike. IBM recently spent $1 billion to acquire Merge Healthcare, a company that handles and processes medical images at more than 7,500 US healthcare sites.

> IBM plans to use its Watson Health Cloud to analyze and cross-reference Merge's medical images against lab results, electronic health records, genomic tests, clinical studies and other health-related data sources amounting to 315 billion data points and 90 million unique records. ... The goal, it said, is to shed new light on current and historical images, electronic health records and data from wearable devices. Armed with the new capabilities, clinicians, for example, will be better able to efficiently identify options for the diagnosis, treatment and monitoring of health conditions such as cancer, stroke and heart disease. Researchers, meanwhile, will gain new insights to aid clinical trial design, monitoring and evaluation. (Noyes 2015)

Individual patients who undergo procedures that include medical imaging know that they are providing data to their doctors for the purpose of diagnosing and treating their conditions or injuries. Those patients signed privacy policies that informed them about other uses of their images, including, for example, the kinds of large-scale research and analysis uses to which Merge and IBM are subsequently putting those images. Eventually, insights derived from analysis of large data sets of images will help medical workers provide better diagnoses and care to patients, perhaps even to the people who provided the images in the first place. However, it is unlikely that the patients specifically understood, at the time they signed the release form, that their images would be leveraged in this fashion. Further, in the age of massive data breaches of medical records, the inclusion of individual records in large collections poses privacy threats to specific patients. Many would argue that the overall positive ends justify the only moderately risky means. Most would agree, however, that medical records—particularly images of specific patients and their conditions and injuries—are particularly sensitive materials.

Information that IBM makes available to the public clearly indicates that the company's approach to, and systems for, big data analysis in the healthcare segment focuses on administration-level systems and decision making. When answering the question "How can Watson Health help you?" IBM provides exemplars targeted at physicians, CIOs, hospital administrators, agency directors, care managers, payers (insurance companies), researchers, public health officials, personal trainers, and nutritionists rather than directed toward patients ("How Can Watson" n.d.).

There are IoT devices and applications that reach consumers-as-individuals, particularly in the 'wearables' product segment, as exercise and health monitors (including devices such as Wi-Fi-enabled pacemakers) proliferate in the marketplace. Additionally, individuals who operate small businesses or other commercial operations, such as farming, benefit from networked data analytics. For example, in France, herd-monitoring technologies notify farmers when cows are in heat or when a calving episode begins ("Cows Can Text" 2012). In Spain, environmental monitors are used to track agricultural parameters such as ambient temperature, humidity, precipitation, wind, or leaf wetness to create statistical models that predict the appearance of particular diseases in vineyards. Monitors can also be used to track vineyard products (grapes and wines) across and throughout the growth/production/delivery pipeline.

> On the one hand, the system allows [us] to monitor vineyard conditions in real-time, being able to predict the appearance of a plague in the next hours/days. This feature allows vineyard technicians to take the measures to minimize the impact of the plague in the vineyard, minimizing time and money lost due to this plague. On the other hand, the system also allows [us] to monitor and control the grape from its beginning to the end user, also called as traceability of the grape. In this way, [the] grape can be monitored in real-time from its plantation to wine manufacturing in the wine cellar. RFID technology allows [one] to accomplish this goal, improving viticulture to a level not known ever before. (Bielsa 2012)

Real-time data analysis can provide meaningful information and feedback from a wide range of devices, monitors, and processes to inform and improve decision making. Situations requiring rapid decision making are particularly opportune for the application of IoT technologies.

For example, IBM works with powerboat racing team owner Nigel Hook to provide real-time data collection, analysis, and representational visualizations that leverage large and diverse data sets, producing actionable information for competitors during races and thereby improving their decision making and effectiveness ("SilverHook Powerboats" 2015).

In 2015, the FTC issued a report about privacy and the IoT. The report was based on the deliberations from meetings held in 2013, as well as on an examination of rapid technological progress in the IoT business segment and consideration of the lack of privacy legislation targeting the IoT. The committee's recommendations included extensive reliance on principles established in the

FIPs. The recommendations encouraged IoT industry leaders and providers to build privacy protections into devices, monitors, and networks rather than merely attempting, later, to add missing and needed protections (Dwoskin 2015). The committee noted that the IoT presents the potential for numerous harms to consumers as well as to its own development/deployment, by

> (1) enabling unauthorized access and misuse of personal information; (2) facilitating attacks on other systems; and (3) creating risks to personal safety. Participants also noted that privacy risks may flow from the collection of personal information, habits, locations, and physical conditions over time. In particular, some panelists noted that companies might use this data to make credit, insurance, and employment decisions. Others noted that perceived risks to privacy and security, even if not realized, could undermine the consumer confidence necessary for the technologies to meet their full potential, and may result in less widespread adoption. (*Internet of Things* 2015: ii)

The report reiterated the importance of the FIPs as well as the need for legislative attention to fundamental privacy legislation that is "flexible and technology-neutral, while [also] providing clear rules of the road for companies about such issues as how to provide choices to consumers about data collection and use practices" (*Internet of Things* 2015: viii).

Recommendations Can Help

Online recommendation systems are not all created equal; many depend on comments contributed by and shared among knowledgeable humans. Although there are privacy issues surrounding user-generated content, recommendations based on human contributions (such as customer comments) of verbal content (written or audio/video recorded comments) are not at issue here. This section presents what is now taken for granted in the electronic communication environment: New media users continually speak with their fingers (behaviors) and they very much like recommendations offered via algorithms. Noting that these recommendations are based on algorithmic calculations does not naïvely ignore the fact that humans program and code algorithms into the operations. Nor does it ignore the fact that humans sometimes play an even stronger role (than merely programming code) in the overall recommendation operation. The focus here, though, is on the relatively simple, obvious, and everyday observation that enormous numbers of individuals freely choose to take advantage of what they perceive as the benefits of algorithmic recommendations.

There are a lot of reasons to criticize the privacy policies of major new media companies such as Google, Amazon, and Netflix. These companies have amassed enormous databases of personal information and are major participants in the data marketplace. That participation is not at issue in this section. People use the Google search engine because it seems to work better than other search engines. The ever-changing Google algorithm often returns search results that meet users' interests and needs. People are pleased when, based on their prior consumption and/or purchase histories, Amazon or Netflix makes a recommendation about additional products or viewing opportunities. In other words, there are many circumstances when users are happy that product/service providers collect massive data sets and use them to add value to consumers' experiences.

The recommendation systems developed and utilized by Netflix stand as a good example. Let us assume, for the sake of this discussion, that Netflix protects its users' privacy by adhering to all of the FIPs, thereby making its system worthy of study (this is probably not the case, in fact, but let's pretend). Examination of its practices shows how and why big data analysis works and produces positive outcomes for consumers.

Netflix collects so much data from its users that it has little need to acquire additional data from the data marketplace. Most of the data that it needs, to produce recommendations, are derived primarily from its members'/users' interactions with the Netflix service and products. Additional insights derived from external sources such as social media (mostly Facebook-based information about what connected friends watch and rate) or publicly available press reports (such as box office performance and critic reviews) refine the data set and analysis.

> We [Netflix] have several billion item ratings from members. And we receive millions of new ratings a day. ... [T]here are many ways to compute popularity. We can compute it over various time ranges. ... Or, we can group members by region or other similarity metrics and compute popularity within that group. We receive several million stream plays each day, which include context such as duration, time of day and device type. Our members add millions of items to their queues each day. Each item in our catalog has rich metadata. ... We know what items we have recommended and where we have shown them, and can look at how that decision has affected the member's actions. We can also observe the member's interactions with the recommendations: scrolls, mouse-overs, clicks, or the time spent on a given page. Social data has become our latest source of personalization features; we can process what connected friends have watched or rated. ... All the data we have mentioned above comes from internal sources. We can also tap into external data to improve our features ... such

as box office performance or critic reviews. ... [T]here are many other features such as demographics, location, language, or temporal data that can be used in our predictive models. (Amatriain and Basilico, "Netflix Recommendations (Part 2)" 2012)

Much of the data Netflix uses for making recommendations are derived from information offered by users during their interactions with the service. Presumably, those users encourage Netflix to use that information in order to provide recommendations for additional programs that the user might like to view. Although Netflix's recommendations take up screen space and thereby limit, to some degree, the range of titles that users can view at one time—a usability issue that limits choice—none of the recommendations are forced on users; no one has to watch the recommended programs. Some users complain about the interface or the degree to which their choices seem to be limited by Netflix's recommendations. However, experienced users understand how to manipulate the interface in ways that moderate the limitations introduced by the interface and the recommendation system. Netflix reports that its users like the recommendation system: "The company estimates that 75 percent of viewer activity is driven by recommendation" (Vanderbilt 2013).

Algorithmic recommendation systems, such as those used by Amazon and Netflix, combine a large number of mathematic/analytical techniques. These systems are generally based on a variety of modifications of two fundamental predictive analysis paradigms: affinity analysis and association rule learning.

"Affinity analysis is a data analysis and data mining technique that discovers co-occurrence relationships among activities performed by (or recorded about) specific individuals or groups" ("Affinity Analysis" 2015). 'Market basket analysis' is a common application of this approach, where the purchasing behavior of specific consumers is identified and recorded, then analyzed via predictive algorithms. "Association rule learning is a method for discovering interesting relations between variables in large databases. It is intended to identify strong rules discovered in databases using some measures of interestingness" ("Association Rule Learning" 2016). Web usage mining makes robust use of association rule learning. Modified combinations of association rule learning and affinity analysis undergird many algorithmic recommendation protocols.

Of course, the most oft-utilized recommendation system on the planet is the big data analytic (and algorithmic) 'secret sauce' that enables the Google search engine. Starting with a collection of data sets (in this case, 'crawled and collected' Web pages on the Internet), then applying the now iconic PageRank algorithm, Google measures the relative importance of sets of hyperlinked documents (and certain elements within those documents) in order to return relevant search results (Brin and Page 1998; "How Search Works" n.d.; "Page-Rank" 2016).

In the world of new media, the thriving, struggling, and sometimes dying of software and platforms such as Netscape, MySpace, America Online, Real Networks, Second Life, and Yahoo, just to mention a few, cautions one against making predictions that seem to be 'sure things.' It is, however, the case that Google touches so much of the World Wide Web, so deeply, that it is very difficult to imagine the Google search engine as anything other than the dominant consumer choice. Such ubiquity indicates consumer acceptance and confidence. This dominance has been achieved in the face of numerous reports in books and in the press warning of the vast amount of personal user data that Google collects. Even consumers who do not study privacy have at least a passing familiarity with the fact that they are giving up information to Google when they search and surf the Web. Nevertheless, one has to suspect that the majority of users very much appreciate the recommendations provided by Google search just as they order and watch the programs Netflix recommends.

Improved Targeting Can Be Good for Both Sides

Hearkening to the early days of targeted marketing/advertising, Crain provides detailed descriptions of Web functionalities—including centralized server network, server log analysis, Web client and Internet domain analysis, protocol-based data such as header analysis, third-party data integration, click-through analysis, user profiling, and HTTP cookies—that implemented surveillance-based ad targeting. Further, he highlights the relationships among corporate pressures for RoI, improvements in targeting and tracking technologies, and implementation of surveillance-based ad targeting.

> [T]he structure of the web technology appropriated by DoubleClick played a fundamental role in the creation of a business model in which every ad served was also an opportunity to gather data about internet users and the efficacy of ad campaigns themselves. In theory, the more ads the company delivered, the more accurately targeted each became. ... Moreover, it set up the technical foundations for the increasingly sophisticated forms of consumer surveillance that followed in the next few years. (Crain 2013: 68–69)

Crain's review highlights the realization that technology could be made to seem 'personal,' thereby improving electronic contacts with potential customers:

> While the web's unruliness contributed to an increasing scarcity of consumer attention to commercial messages, marketers and infra[structure] providers also saw potential in utilizing internet technology to learn about consumers in order to better tailor marketing interactions. The notion of personalized ads represented a new twist on a late twentieth-century understanding of customer relationship management (CRM). ... [M]arketing gurus such as Don Peppers and Martha Rogers promoted the concept of "one-to-one" or "relationship" marketing as a method of maintaining engagement with a company's most important customers and prospects. A mantra of CRM was that a majority of profits stem from the repeat business of a select group of loyal customers. As such, marketers needed to cultivate relationships with this group and seek out others who would "act like best customers once they are brought into the fold." The goal was to engage such targets in conversation—to speak to consumers, but also to watch and listen. (Crain 2013: 156)

Beginning roughly in 1998, targeted marketing, based on user data, became the principle change agent in online industries.

> [W]hat changed in this second phase of surveillance was the extent to which third party data was utilized and the nature of the information itself. A major trend was the integration of increasingly sensitive forms of consumer information. ... Surveillance infrastructure development was driven by marketers' overarching needs to bring the internet into the marketing complex and, to the greatest extent possible, maximize its utility for purposes of selling. (Crain 2013: 162, 178)

Google's purchase, in April 2007, of DoubleClick ($3.1 billion) sealed the deal, combining DoubleClick's robust tools and infrastructure for surveillance-based ad targeting with Google's dominance of both search and, eventually, advertising.

There is widespread disagreement concerning consumers' understanding about, and/or appreciation and acceptance of, targeted marketing and advertising. Neither the hyperbolic and self-serving claims issued by industry proponents as to the values derived by consumers from targeted materials nor the equally contentious claims by privacy advocates that all targeting is odious are realistic or helpful. Positions articulated on both sides contain truthful aspects. Clearly, this book tends to side with privacy advocates and shares many of their concerns. However, one cannot dismiss out of hand all aspects of the industry claims about consumer benefits. At least in some respects, consumers themselves recognize, utilize, and prefer certain benefits provided by targeted marketing and advertising.

In 2013, a public opinion poll commissioned by the Digital Advertising Alliance indicated that a low percentage (4%) of respondents placed targeted marketing and advertising high among a list of their negative concerns about the Internet. Further, the respondents stated overwhelmingly that they prefer free content, and content that is supported by advertising, to paid services (Bachman 2013). According to the polling firm Zogby Analytics,

> What the poll makes clear is that consumers prefer ads that reflect their particular interests, which is precisely what interest-based advertising was created to provide. ... The poll also demonstrates that Americans' privacy concerns are rightly focused on real threats like malware and identity theft, and not on an industry that follows rigorous, enforceable guidelines for data collection and use. ... More than 90 percent of Americans polled said that free content was important to the overall value of the Internet, and more than 60 percent said it was "extremely" important. Similarly, more than 75 percent of poll respondents said they prefer content (like news, blogs and entertainment sites) to remain free and supported by advertising, compared to fewer than 10 percent who said they'd rather pay for ad-free content. ("Poll: Americans" 2013)

Turow (2011) and others criticize the study as industry-driven fodder, its results manipulated via loaded questions that respondents did not understand (Bachman 2013). That criticism is probably accurate in a strictly technical sense. The problem, however, is that most users of digital technologies behave roughly in line with the findings of the survey rather than following the democracy-protecting advice of privacy advocates. As noted repeatedly in this book, most operations of the data marketplace are secret, hidden in plain view, overly technical, and in a wide variety of ways, difficult for everyday citizens to understand and appreciate. The complexities forgive neither the targeters nor the targeted. Regardless, most people prefer that the Internet remain seemingly free and generally understand that targeted advertising is part of the price that must be paid.

Consumers are, however, generally conflicted over their allegiance to factors they suspect are in play. Jack Marshall reports in the *Wall Street Journal* that

> while some consumers see benefit in tailored ads, most don't want their information to be used to tailor them. ... Data from real-time research company Qriously tells a similar story. ... 54% of respondents said they prefer relevant ads to irrelevant ones. But ask a question about targeting and it's a slightly different story. Only 48% said they preferred "targeted" ads over "non-targeted" ones. (Marshall 2014)

It is not difficult to notice the ways that consumers have adjusted to realities on the Internet.

> In the past five years, online advertising has become incredibly intelligent. We are now at the stage where ads can be served based on what consumers are sharing and talking about. Sharing has become something of a phenomenon and can come in all forms, whether it be a tweet, a shortened URL, even an email telling someone to look at a link. Advertisers are increasingly able to build profiles of people, based on their interests and what they are sharing across the Open Web, and serving relevant ads accordingly at to scale. Consumers have reacted well. They understand that they are going to be served ads online these days—it's what makes the internet tick—so they may as well be useful. (Staines 2014)

Like it or not,

> [m]ining data is the weapon of choice for the Mad Men of our day because of its rapier-like accuracy and striking at the heart of our desires. ... The main driver of profit on the Internet is neither subscription income nor general space ads. ... What is truly revolutionary about the Internet is both the completeness of the profiles of the individual customers being delivered to advertisers and, most importantly, the ability to determine via click-through whether contact with the customer results in an actual sale. ... [T]he unique value of Internet data mining is this ability to fine tune outreach and manipulation of individual desire based on the most intimate mapping of personal taste, fears, and aspirations. (Scheer 2015: 60, 68)

According to Tanner, "the best businesses give consumers a choice whether or not to share their data, and offer benefits in return" (2014: 171). This aspect of free choice is unfortunately not part of the business models of most online participants in the data marketplace. One cannot rule out the potential that online entities will strike the right balance among transparency, data collection, consumer awareness and permission, and provision of incentives, thereby optimizing advantages for all. To that end, is useful to realize that certain amounts and kinds of tracking and recommendations might provide benefits to both sides in the economic exchange.

The Usability and Functionality Lies

So many variations of the 'usability and functionality lies' are used on the Internet that merely listing them would fill a book chapter. I will not attempt an exhaustive list. McChesney claims that "this may be the great Achilles' heel of the Internet under capitalism: the money comes from surreptitiously violating any known understanding of privacy. The business model for Google and Facebook, and to a certain extent for all Internet firms ... requires a magic formula to appear in which some method of violating privacy and dignity becomes acceptable" (2013: 151). I will represent the usability and functionality lie by presenting material taken from the privacy policies of a popular website: CNN.com. My comments are inserted [in bold within brackets].

IV. Cookies and Other User and Ad-Targeting Technologies.

We use cookies and other technologies both to provide our Services to you and to advertise to you. We also may work with Partners to help them advertise to you when you visit other websites or mobile applications, and to help them perform user analytics. These technologies may also be used to analyze how our Users interact with advertising on our Services and elsewhere, and more generally, to learn more about our Users and what services or offers you might prefer to receive. We describe some of these technologies below.

1. Cookies. To enhance your online experience, we and our Partners use "cookies", "web beacons" or other tracking technologies.
[This is perhaps the biggest lie of all. As acknowledged in the preceding paragraph, the company uses these technologies to provide the product or service in a fashion that suits it and that suits its connections to advertisers. The only connection to improving users' online experiences is that if one disables cookies or beacons, the website does not function properly. As noted below, this is a feature that the company purposefully imposes on the website and user. In other words, "Take it or leave it; it is our way or the highway."] Cookies are text files placed in your computer's browser to store your preferences. We use cookies or other tracking technologies to understand Service and Internet usage and to improve or customize the products, content, offerings, services or advertisements on our Services. For example, we may use cookies to personalize your experience at our Services (e.g., to recognize you by name when you return to a Service), [Some users might like being recognized by name; others might not consider this feature an enhancement to their online experience.] save your password in password-protected areas, and enable you to use shopping carts on our Services. We also may use cookies or other tracking technologies to help us offer you products, content, offerings or services that may be of interest to you and to deliver relevant advertising when you visit this Service, a Turner Affiliate's Service, or when you visit other websites or applications. [Only some users will consider being offered products or services that "may be of interest to them" as an enhancement of their online experience.] We or a third party platform with whom we work may place or recognize a unique cookie on your browser to enable you to receive customized content, offers, services or advertisements on our Services or other sites. [These third parties and/or partners are never identified.] These cookies contain no information intended to identify you personally. The cookies may be associated with de-identified demographic or other data linked to or derived from data you voluntarily have submitted to us (e.g., your email address) that we may share with a service provider solely in hashed, non-human readable form.

[This is the second biggest usability and functionality lie; later in this policy, CNN admits as much and details the various ways in which this factor is a lie. The presentation of tracking technologies as de-identified is a lie. In cases such as this, it is told as a partial truth: The cookies and beacons referred to in this paragraph do not contain information that personally identifies the source of the information. However, the data marketplace enables virtually all participants, especially and including CNN in this case, to identify the sources of tracking data. That is, after all, the point: to advertise to specific people who are interested in specific products.]

We, our third party service providers, advertisers, advertising networks and platforms, agencies, or our Partners also may use cookies or other tracking technologies to manage and measure the performance of advertisements displayed on or delivered by or through the Turner Network and/or other networks or Services. This also helps us, our service providers and Partners provide more relevant advertising.

2. Syncing Cookies and Identifiers. We may work with our Partners (for instance, third party ad platforms) to synchronize unique, anonymous identifiers (such as those associated with cookies) in order to match our Partners' uniquely coded user identifiers to our own. We may do this, for instance, to enhance data points about a particular unique browser or device, and thus enable us or others to send ads that are more relevant, match Users to their likely product interests, or better synchronize, cap, or optimize advertising.
[As noted above, the "de-identified data" lie is hereby outed.]

3. Locally Stored Objects. Services on the Turner Network may employ locally stored objects ("LSOs") and other client side storage tracking technologies in certain situations where they help to provide a better user experience, such as to remember settings, preferences and usage similar to browser cookies, or in order to target or help our Partners target ads, analyze ad performance, or perform user, website or market analytics. [Once again, the term "to provide a better user experience" is used fairly loosely. Most of the functions listed here refer to helping CNN better target ads at users.] For LSOs utilized by Adobe Flash you can access Flash management tools from Adobe's website. In addition, some, but not all browsers, provide the ability to remove LSOs, sometimes within cookie and privacy settings.

4. Disabling Cookies. Most web browsers are set up to accept cookies. You may be able to set your browser to warn you before accepting certain cookies or to refuse certain cookies. However, if you disable the use of cookies in your web browser, some features of the Services may be difficult to use or inoperable.
We may work with certain third-party companies that use techniques other than HTTP cookies to recognize your computer or device and/or to collect and record information about your web surfing activity, including those integrated with our Services. Please keep in mind that your web browser may not permit you to block the

use of these techniques, and those browser settings that block conventional cookies may have no effect on such techniques. To learn more about Interest-Based Advertising or to opt-out of this type of advertising by those third parties that are members of DAA's opt-out program, please go to http://www.aboutads.info. [**The section is an example of "hiding in plain sight." Having just warned users that turning off cookies and beacons will, more often than not, disable website functions, CNN now tells users that CNN utilizes technologies that attempt to cancel out user efforts to evade cookies and beacons. In other words, "We are doing this for your own good and, even if you don't appreciate that, we will do everything we can to collect the information anyway."**]

5. Web Beacons. We and our Partners may also use "web beacons" or clear GIFs, or similar technologies, which are small pieces of code placed on a Service or in an email, to monitor the behavior and collect data about the visitors viewing a Service or email. For example, web beacons may be used to count the users who visit a web page or to deliver a cookie to the browser of a visitor viewing that Service. Web beacons may also be used to provide information on the effectiveness of our email campaigns (e.g., open rates, clicks, forwards, etc.). [**As noted in the preceding paragraph, some surveillance technologies that collect data cannot be turned off via common practices. Turning off cookies does not affect the ways that beacons operate.**]

6. Mobile Device Identifiers and SDKs. We also sometimes use, or partner with publishers, publisher-facing, or app developer platforms that use mobile Software Development Kits ("SDKs"), or use an SDK with a mobile app that we offer, to collect Information, such as mobile identifiers (e.g., IDFAs and Android Advertising IDs), and Information connected to how mobile devices interact with our Services and those using our Services. A mobile SDK is the mobile app version of a web beacon (see "Web Beacons" above). The SDK is a bit of computer code that app developers can include in their apps to enable ads to be shown, data to be collected, and related services or analytics to be performed. We may use this technology to deliver or help our Partners deliver certain advertising through mobile applications and browsers based on information associated with your mobile device. [**This section contains a significant lie by omission. Data from mobile devices is not de-identified; there is no way to collect information from a user's specific mobile phone without also collecting identity-specific information tied to that cellular account. This aspect plays a significant role in the ability of Web services to 'on board' data about users by connecting online with offline behavior**] If you'd like to opt-out from having ads tailored to you in this way on your mobile device, please follow the instructions in the "Your Ad Choices" section below.

By visiting the Service, whether as a registered user or otherwise, you acknowledge, and agree that you are giving us your consent to track your activities and your use of the Service through the technologies described above, as well as similar technologies developed in the future, and that we may use such tracking technologies in the emails we send to you. [**This is the third major usability and functionality lie: that by merely using a website, users consent to tracking and the uses to which those data are put.** As noted previously, privacy policies and ToS are insidious with regard to this feature by establishing wrap contracts—odious contracts of adhesion—as valid agreements of assent. They are not. They are more often representations of combinations of users' ignorance and situational exigence; users have neither the time nor the patience for the intricacies of complex and long privacy policies and/or ToS. Neither those states nor use of the site indicates assent.]

Our unaffiliated Partners and third parties may themselves set and access their own tracking technologies when you visit our Services and they may have access to information about you and your online activities over time and across different websites or applications when you use the Service. Their use of such tracking technologies is not in our control and is subject to their own privacy policies. [**This is the fourth major usability and functionality lie: Here CNN claims that it cannot control what third parties do on its websites. This lie is a misstatement.** Readers are supposed to understand this to mean that CNN cannot control the use to which third parties put user data. However, here CNN claims that it cannot control third-party tracking on its websites. This is simply not true. What is true is that if CNN wants to allow its affiliated partners to place third-party trackers on the CNN website, it is unable to keep unaffiliated parties from negotiating third-party tracking on the site. If CNN turned off third-party cookie functionality, it would control the access that unaffiliated third parties have to user data.] ("Cookies and Other" 2015)

For the World Wide Web to function the way it currently does—for it to produce massive amounts of profit for entities within the data marketplace—the companies that provide websites, and the products and services found there, must participate in a symbolic, rhetorical, persuasive, ideological, and untruthful communication exchange with users. In ways that are self-serving and purposefully surreptitious, providers must convince users that technological actions that track them and collect personal information about them, for the purposes of marketing and advertising, are done for the benefit of those users.

Usually, most users trust that Internet companies have their best interests in mind and at heart. Further, Internet companies must convince consumers that this is the only way the Web will work, the only way that they—the providers—can do their jobs. 'This is just the way the Web works' is a common sentiment. Readers who understand theoretical concepts surrounding media determinism know that such claims are strategic misdirection. Although it is the case that mediums may have preferences, technological mediums do not 'do' anything on their own. The World Wide Web is programmed by human beings carrying out particular actions by way of code and software delivered through networks and devices (which also use code and software). When new media technologies mislead or deceive users, it is because the people who develop and deliver them use deceptive techniques to further their business objectives. They may feel as though the subterfuge is justified; they may feel as though the products and services they provide give consumers more benefits than damages via costs. But the game currently being played by most companies and providers is inherently dishonest.

Perhaps worse, constructing the Internet in this fashion encourages even nonprofit entities to participate in most of the same behaviors. Hospitals, social service agencies, medical practices, public/governmental entities, church groups, schools at all levels—in short, entities that want to draw and communicate with an audience on the Internet are faced with an environment that virtually requires them to use invasive surveillance technologies in order to make their websites comparable to the commercial sites that users are used to. The result is a race to the bottom of the privacy pool.

Unreliability via the Marketplace

Marketing researcher and curricula developer Greg Hamilton notes that "[t]argeted marketing, the practice of aiming marketing collateral at specific prospects or customers, has become so prolific that it is one of the largest tools in the modern marketer's toolkit" (Hamilton 2015). The process of targeted marketing is complex and requires good data, sound analysis, and adept and intuitive application to specific business challenges and opportunities. Targeted marketing can go wrong in a number of places along its value chain. Although touted as a quantum-leap improvement over earlier marketing and advertising techniques, targeted marketing informed by big data continues to feature numerous shortcomings, some of which suggest that the personal information that consumers give up is not always rewarded with improved marketing and advertising outcomes.

Inaccuracy of data is a significant shortcoming of targeted marketing by way of big data analysis. In the process of examining the data marketplace and privacy, Julia Angwin completed an audit of the personal information (about herself) that she could access (see *Dragnet* 2014: chapter 6). Putting aside for the moment the fact that she had difficulty retrieving information from many of the companies she contacted, Angwin also reported particularly troublesome results related to inaccuracies in many of the reports she was able to retrieve.

Writing about her experience with the 'free' annual credit reports that must be provided (under federal law) by the big three credit reporting agencies—TransUnion, Experian, and Equifax—she noted that "the latest review of credit report accuracy by the Federal Trade Commission determined that 26% of people found at least one significant error on at least one of their three reports" (Angwin, *Dragnet* 2014: 90). Angwin's credit agency report results were even less impressive and less accurate than that, as she found a number of errors and inaccuracies in the reports. She lamented even less accurate outcomes from information collected by firms in the data-scoring business, outfits that performed so poorly for her audit that she opted out of their services.

The report she received from Acxiom, one of the largest data brokers in the world, featured remarkably poor demographic information about Angwin. Her report from another large data broker, Datalogix, contained several incorrect inferences and categorizations about her, based on the data it had matched to her profile. Overall, Angwin was nearly stunned by the amount of inaccuracies in much of the information contained in her digital dossier. Much of it was correct; however, her audit uncovered reports that contained many inaccuracies.

Angwin worked much harder at her information audit than do most people. Generally, people are unable to ascertain or fix inaccuracies in private information contained within the data marketplace. Inaccurate information can damage credit scores, reputations, and opportunities, and of course, user inability to ascertain or correct information is not in accord with the second and fourth FIPs.

Given the amounts of data produced by contemporary data collection techniques, errors are also generated by analysts armed with inadequate analytical tools. The research outfit Gartner attempts to capitalize on these challenges and shortcomings in its promotional materials. In doing so, it highlights some of the most significant shortcomings of targeted marketing and advertising that relies on big data:

> Digital technologies are producing mountains of data. Digital marketers know there's gold in those mountains, but most lack the treasure maps, tools and processes to extract it. The old techniques of market research and data analysis no longer work because the variety, velocity and volume of information have reached unprecedented levels. ... Analytical skills are falling short of demand. New technologies that generate data often create information silos that digital marketers struggle to break down. (Kihn 2014: 2, 3)

Perhaps most disconcerting of all in this age of matching data to advertising targets, "B2C companies collect data about vast numbers of people, many of whom are anonymous" (Kihn 2014: 3). Much sophisticated data analysis is still based on data from anonymous sources and yet is extrapolated to specific targets.

Hamilton highlights another aspect of targeted marketing inaccuracies: directing targeted marketing appeals toward the wrong audiences. Although the goals of targeted marketing seek to match advertising with particular consumers, the realities of the current state of affairs find that marketing efforts very often fall short of the ability to accurately target specific individuals. As a result, he cautions, "We as marketers must recognize that, while our intended prospects and customers are included in our target group, by their very nature marketing messages will also be consumed by those who are not a part of that group. It is never customer-centric to offend another group while targeting your desired customer segment" (Hamilton 2015).

Following on this trend, Teresa Fernandes and Ana Calamite report that targeting and specialization can actually work against the economic goals of marketers and advertisers by leading to negative responses from their own customer base.

> Questionable marketing tactics to attract and retain customers, such as differential treatment, favouritism and data use, raise concerns about the fairness consumers experience from retailers. … We argue that existing customers may perceive offers that favour new customers as unfair, prompting negative attitudinal and behavioural consequences, in particular with regard to satisfaction, trust and intention to repurchase. … Different unfairness perceptions result in different consequences in terms of satisfaction and trust which will then mediate the impact on repurchase intentions. In the case of trust, we further concluded that differential pricing favouring new customers may jeopardize the trust of both existing and new clients, since a violation of society norms is a source of distrust. (Fernandes and Calamite 2016: 42)

A significant, and very worrisome, trend connected to inaccuracies in targeted marketing relates to industry aspirations, toward which industry players currently strive, and then fall short, thereby damaging consumers. One indication of this trend is represented by comments from Google's CEO Eric Schmidt. In an oft-cited *Wall Street Journal* interview, Schmidt told Holman Jenkins:

> I actually think most people don't want Google to answer their questions. … They want Google to tell them what they should be doing next. … The power of individual targeting—the technology will be so good it will be very hard for people to watch or

consume something that has not in some sense been tailored for them. (quoted in Jenkins 2010)

In more technical language, Schmidt was referring to industry-wide moves toward sentiment analysis and predictive analysis. "The goal [of sentiment analysis] is not to describe but to affect and effect—to stimulate word-of-mouth, to promote engagement and, in some cases, to thwart it. ... The goal of predictive analytics is both pre-emptive and productive: to manage risks before they emerge or become serious all the same time maximizing sales. Such approaches, in other words, seek to integrate possible futures into present behavior" (Andrejevic 2013: 57).

Harcourt notes that the dangerous aspect in this procedure is that it fosters efforts to "find our *doppelgänger* and then model that person's behavior to know ours, and vice versa, in the process shaping each other's desires reciprocally":

> It is not just feedback but feedforward: to look forward by looking backward, back and forth, so that someone is always in front of the other, behaving in a way that the other will want next. You never double back; you are always at the same time leading and following, abuzz with action, shaping and being shaped, influencing and being influenced. ... This new logic—this a doppelgänger logic—differs from the forms of rationality that governed earlier. Today, the capabilities and logic evolved toward the ideal of the perfect match. Yet there is one thing that runs through these different logics: data control. In the digital age, this means the control of all our intimate information, wants, and desires. (Harcourt 2015: 145, 146)

When judging the dangers involved in using big data and targeted marketing/advertising to reach consumers, an analogy with voice recognition software is appropriate. As is the case with targeted marketing and advertising informed by big data analysis, voice recognition software has improved exponentially over the course of its development in the past two decades (Oremus 2014). Particularly when used within domains with limited, specific, and specialized vocabularies (for example, medical transcriptions), voice recognition software can be very accurate.

However, in everyday, common usage, the software can still make significant numbers of mistakes, perhaps reaching 92% to 94% accuracy even under relatively controlled dictation conditions. Applying this analogy to predictive and sentiment analytics, it is not difficult to imagine the havoc that can be played when personal data are used to influence 6% to 8% of the wrong people or to influence the right people toward the wrong outcomes 6% to 8% of the time.

These circumstances find consumers in a no-win situation. When targeted marketing 'gets it right,' consumers face digital market manipulation (Calo 2014). When targeted marketing 'gets it wrong,' consumers not only receive marketing and advertising appeals that are inappropriate for them, they must fend off self-consciously aggressive attempts to change their affects and effect their behavior as the marketplace seeks to mold consumers into sought-after targets.

Data Abuse via the Marketplace

Turning citizens into "targets or waste" (Turow 2011: 88–111) has serious social and moral consequences. Audience segmentation, sentient and predictive analysis, brought to highly effective results via targeted marketing and advertising, not only segregate and discriminate; these practices also abuse the very humanity of citizens.

Vincent Miller draws our attention to what the process focuses on and what it ignores:

> The problem lies in how such presences [ours] are turned into abstracted forms of 'data'. As data, this virtual matter is conceived of as 'information about' beings as opposed to 'the matter of being' in contemporary environments. This allows aspects of the contemporary self (i.e. data about ourselves) to be treated as commodities, not as meaningful components of the self but as a series of potentially useful or valuable objects (data to be used and sold). This 'matter as data' carries with it no ethical weight, and thus the handling of personal data is, therefore, largely free from any kind of ethical or moral responsibility that one might have towards a physically co-present human. This encourages the rampant collection of data, invasions of privacy and the spread of personal information. (Miller 2016: 8)

Advocates of targeted marketing are likely to read this passage and say, "Yes, he's right, we collect a lot of data about people." Such a reading misses the point. The point is that targeted marketing does not collect data about people as people; it collects information about people as data points. People working in the data marketplace might well claim that collecting data is an amoral activity but that as long as they do not misuse the data the practice is moral. Quite to the contrary, Miller's point is that treating people as data points is abusive by definition because it ignores their human agencies, including, for example, the abilities they should have to control their own private information. Further, collecting data with this mindset leads to misuses of the information. One of the ways we can verify that Miller's concerns are well founded is to reflect on the reality that very few collectors and miners of data follow the FIPs.

One of the most ubiquitously repeated defenses for data collection and targeted marketing is that the marketing and advertising purposes to which the data are put are fairly trivial. "Oh, it's just for advertising" is the common refrain. Unfortunately, that is not the case—even within the commercial market, let alone considering the uses to which the government puts information that it acquires from entities within the data marketplace.

Turow's explanation of consumers as target or trash highlights, yet understates, the issue:

> Those considered waste are ignored or shunted to other products the marketers deem more relevant to their tastes or income. Those considered targets are further evaluated in the light of the information that companies store and trade about their demographic profiles, beliefs, and lifestyles. The targets receive different messages and possibly discounts depending on those profiles. (Turow 2011: 88)

The outcome Turow presents is discrimination in the provision of discount coupons and other differential pricing strategies. If the damages were limited to these activities one might agree with the proponents who claim that these are relatively trivial advertising and marketing strategies. Unfortunately, the stakes are much higher and the outcomes much more dire.

Matthew Crain details much more serious considerations:

> the root of the issue, [which] is a hardening of social class along lines of difference whereby marketing practices reproduce social discrimination and economic exploitation on the internet and beyond. In this scenario, the marketing complex is doing more than just sorting profiles according to differential consumer typologies. It is creating a system that when functioning optimally catalogues individuals as either valuable or worthless to highly specific processes of profit maximization. (Crain 2013: 185)

This is the kind of sorting of targets and trash that produces real and unacceptable discriminatory practices. As noted by Stephen Cory Robinson, "consumers can experience weblining, the practice of denying someone access and service based on their online presence[ ]. Consumers with an undesirable profile may not be offered the same goods and services as consumers who fit a profile deemed more desirable by a marketer or organization [ ]" (Robinson 2015 [citations omitted]). The 2014 Presidential privacy report noted:

> Just as neighborhoods can serve as a proxy for racial or ethnic identity, there are new worries that big data technologies could be used to "digitally redline" unwanted groups, either as customers, employees, tenants, or recipients of credit. A significant finding of this report is that big data could enable new forms of discrimination and predatory practices. (*Big Data* 2014: 53)

While acknowledging relationships between data surveillance and ordinary marketing, Neil Richards warns that "the power of sorting can plead imperceptibly into the power of discrimination":

> A coupon for a frequent shopper might seem innocuous, but consider the power to offer shorter airport security lines (and less onerous procedures) to rich frequent fliers, or to discriminate against customers or citizens on the basis of wealth, geography, gender, race, or ethnicity. The power to treat people differently is a dangerous one, as our many legal rules in the areas of fair credit, civil rights, and constitutional law recognize. Surveillance, especially when fueled by Big Data, puts pressure on those laws and threatens to upend the basic power balance on which our consumer protection and constitutional laws operate. As Professor Danielle Citron argues, algorithmic decision making based on data raises issues of "technological due process." [ ] The sorting power of surveillance only raises the stakes of these issues. After all, what sociologists call "sorting" has many other names in the law, with "profiling" and "discrimination" being just two of them. (Richards 2013: 1 [citation omitted])

Sylvia Peacock documents the process by which "the data of mature, well-off and educated men and women are captured by the web tracking technologies of online companies, to an increasing degree" (2015: 1). Since marketing firms prefer these consumers, data collection algorithms focus on collecting information about them. Peacock notes that the preference marketers have for these data (these customers) makes it unlikely that the industry will self-regulate its predilections for collecting them. As a result, broader data generated within social media are not collected, leading to what Peacock refers to as the *tragedy of the commons*.

> Intense Internet users are the ones who are for the most part exposed to a now common but largely unspecified personal data extraction while offliners remain untouched. ... [It is] the personal data of middle-aged, well-educated, wealthier segments of the Canadian population that are tracked and with high probability analysed for marketing purposes. ... In due time, the overuse of the Internet commons, and in particular social media, for the purpose of personal user data extraction might lead to an underuse of its connective possibilities, as online users become increasingly disenchanted. (Peacock 2015: 5)

Seeta Peña Gangadharan notes that discrimination based on digitally automated systems exacerbates inequities already in place:

> These systems run on the data produced in our daily digital meanderings and on algorithms trained to identify patterns among different data points. The result of these computerized calculations include predictions of our future behavior, recommendations for the purchase of one product or another, advice that we modify our behavior, feedback and adjustments to the operation of computer controlled systems, and more. From White House officials to civil rights advocates to "quants" and "techies," many have begun to question the power of algorithmically driven systems to categorize, nudge, prime, and differentially treat people in ways that can exacerbate social, economic, and racial inequities. (Gangadharan 2014: 2)

General concerns are one thing—and can be dismissed as hypothetical hysteria; specific harms are another thing. Tim Libert documents harms that come from data mining practices related to targeted marketing and advertising. For example, he found that over 90% of the 80,000 health-related Web pages he examined "leaked data" that would enable data miners to identify specific patients and their maladies (Libert 2014: 10). Most commercial entities that collect such information are not subject to HIPAA privacy restrictions. "To take one example, as of April 2014, a user visiting the CDC website would have had their browsing information transmitted to Google, Facebook, Pinterest, and Twitter" (Libert 2014: 12–13).

Virginia Eubanks illustrates various ways that using machine-driven decision making can join with data collection and targeted marketing to produce a wide range of discriminatory practices:

> Early big data systems were built on a specific understanding of what constitutes discrimination: personal bias. Discrimination can only be individual and intentional, a caseworker applying welfare eligibility rules more strictly to African American mothers, a police officer finding white citizens somehow less suspicious. By contrast, computers judge "fairly," applying rules to each case consistently and without prejudice. According to legal scholar Danielle Keats Citron, digital systems today go beyond applying procedural rules to individual cases; instead, these systems are primary decision-makers in public policy. [ ] A computer system can terminate your food stamps, exclude you from air travel, purge you from the voter rolls, or decide if you are likely to commit a crime in the future. This presents significant challenges to due process, procedural safeguards of administrative law, and equal protection assurances. How can you prove a discrimination case against a computer? Can due process be violated if an automated decision-making system is simply running code? (Eubanks 2014: 50–51 [citation omitted])

Most readers are familiar with—in fact, may participate in—one of the most common discriminatory practices begat by big data analysis and targeted marketing. Professionalized social networks such as LinkedIn support discriminatory practices in job-seeking markets. While equal opportunity employers cannot discriminate based on race, sex, creed, religion, color, and national origin (among other categories), employers are free to discriminate based on social networks.

Data analytic recruitment filtering may put applicants who do not belong to LinkedIn at a disadvantage compared with those who do. Applicants whose profiles display particular people as part of their network can be preferred over those whose display other people.

> Employers use LinkedIn and other social network sites to determine "cultural fit," including whether or not a candidate knows people already known to the company. This process rewards individuals on the basis of their networks, leading companies to hire people who are more likely to "fit the profile" of their existing employees—to the detriment of people who have historically been excluded from employment opportunities. While hiring on the basis of personal connection is by no means new, it takes on new significance when it becomes automated and occurs at large scale. What's at stake in employment goes beyond the public articulation of personal contacts. While LinkedIn is a common tool for recruiting and reviewing potential professional employees, fewer companies using it for hiring manual or service labor. For companies who receive thousands of applicants per opening—especially those who are hiring minimum wage or low-skill labor—manually sorting through applications is extremely time consuming. As a result, applicant tracking and screening software is increasingly used to filter candidates computationally, especially at large enterprises. (boyd, Levy, and Marwick 2014: 53–54)

In short, data collection and the analysis that produces targeted marketing and advertising carry with them wide-ranging potential and actual damages to the people whose private information is traded within the marketplace.

You Can Touch This Because I Can't Protect It

As indicated in chapter 1, characterizing 2015 as "the year of the breach" indicates the severity of problems with data security. Without significant interventions that solve a wide variety of shortcomings, 2016 and beyond do not promise improved data security. In the absence of adequate security, it is very difficult to justify the collection of massive amounts of personal data, ostensibly for marketing and advertising. Data breaches put that valuable personal data to nefarious uses, costing both individual providers of the data (users) and the economy as a whole massive amounts of value (in time, effort, emotional distress, and money).

To begin, one has to admit that the explosion of targeted marketing and advertising driven by big data analysis raises the likelihood of data breaches simply based on the exponential increase in data marketplace volume. Further, the massive increases in computational resources and network infrastructure threaten scale-based failures:

> Such an infrastructure comprises of thousands of computer and storage nodes that are interconnected by massive network fabrics, each of them having their own hardware and firmware stacks, with layers of software stacks for operating systems, network protocols, schedulers and application programs. The scale of such an infrastructure has made possible service that has been unimaginable only a few years ago, but has the downside of severe losses in case of failure. (Naeimi et al. 2013)

Naeimi et al. (2013) propose a variety of cloud-based solutions for these weaknesses—a solution set that might solve the technological weaknesses (joining machine and cloud-based resources more efficiently) while raising additional data security concerns (by spreading data across increasing numbers of systems and networks).

The rate of breaches in 2016 has not slowed significantly from the record-setting pace of 2015. By March 22, 2016, the Identity Theft Resource Center (ITRC) had reported 177 breaches of over 4.6 million records (*ITRC Data Breach* 2016). The pace amounts to, roughly, 62 breaches a month compared to approximately 65 per month the previous year (*ITRC Breach Statistics 2005–2015*).

Data giant Experian predicts five data breach trends for 2016:

- The EMV Chip and PIN liability shift will not stop payment breaches.
- Big healthcare hacks will make the headlines but small breaches will cause the most damage.
- Cyber conflicts between countries will leave consumers and businesses as collateral damage.
- 2016 U.S. presidential candidates and campaigns will be attractive hacking targets.
- Hacktivism will make a comeback. (*Third Annual* 2016: 2)

It is nearly impossible to find a data analysis or security expert who believes that data breaches will lessen in either the short term or foreseeable future. Most experts predict increases in a wide variety of cybercrime, not limited to, but probably led by, identity theft. Analysts focus particularly on the developments surrounding the IoT "for signs that cyber-thieves are turning their sights to the billions of devices that are fast becoming part of our everyday computing environment" ("Experts Predict" 2015).

Experts identify increased efforts on the part of cyber criminals to destroy networks and infrastructure as well as to collect information as both primary and collateral damage (Thompson n.d.). Further, cyber criminals will use techniques such as 'leapfrogging' to access information contained within networks, by way of breaking down barriers at a central node. This strategy becomes especially important as the IoT routes private information through extensive collections of information-bearing devices, monitors, and data centers. Cloud-based storage and computing are particularly vulnerable to 'Trojan horse'–type attacks such as the newly popular 'Dyre' approach that uses the cloud both to induce users to download the malware as well as to then collect data from unsuspecting clients of the cloud-based operation (Masters 2016; Narayan 2014).

Predictions about preferred methods of illicitly acquiring private data suggest that particularly forceful and negative strategies will increase. For example, one commentator suggested that 2016 will be known as "extortapalooza" in that ransomware will continue as a preferred strategy for increasing user's financial remuneration to breachers (Wendy Nather, quoted in Masters 2016). In 2015 alone, ransomware/malware known as CryptoWall and CryptoLocker extracted more than $23 million from victims (Gallagher, "FBI Says" 2015; Ward 2014).

Although making predictions about new media is risky, data-based extrapolations for data breach frequency can be downright frightening. Spencer Wheatley, Thomas Maillart, and Didier Sornette propose that "Statistical modeling of breach data from 2000 through 2015 provides insights into this risk" (2016: 1). Their findings suggest that "the total amount of breached information is expected to double from two to four billion items within the next five years, eclipsing the population of users of the Internet. This massive and uncontrolled dissemination of personal identities raises fundamental concerns about privacy" (Wheatley, Maillart, and Sornette 2016: 1). Just as electronic mail has been largely overwhelmed by spam, we face the risk of having data breaches overwhelm the safe and efficient use of computational networked communication.

Having reviewed the state of affairs, in light of harms and the five FIPs, we will turn our attention, in chapter 7, to recommendations for solutions. Suffice to say, the scope of the crisis presents daunting challenges to potential solutions.

· 7 ·

RECOMMENDATIONS

It is what it is.
  —mantra spoken by numerous Cisco Inc. employees, San Jose, CA[1]

If I could change the world.
  —"Change the World," Tommy Sims, Gordon Kennedy, and Wayne Kirkpatrick
    (famously recorded by Eric Clapton)

Let us begin with a series of stipulations.

First: The Internet is not going to go away and 'it is what it is.' The business model based on banner advertising that supported the static, non-interactive, Web 1.0 Internet did not produce sufficient revenue to drive profit, investment, and runaway innovation. The business models based on targeted marketing and advertising that support the interactive, Web 2.0 Internet produce significant revenues that support the continued development and improvement of the Internet. Further, many of the operations that utilize the Internet can claim a variety of legal protections afforded to businesses and news gathering/disseminating operations that insulate them from wanton and reckless government intervention. Similarly, none of the following business segments are going to fold up their tents and go away: credit card providers, mobile phone services, corporate loyalty card issuers, or manufacturers of IoT devices and network infrastructure operations for the IoT. These companies 'are what they are' and data about customers are important aspects of, and contributors to, their business models.

Second: Targeted marketing and advertising currently depend on the collection of personal data in ways that require overly invasive and inadequately controlled privacy intrusions. Nevertheless, many Americans do not interpret these activities as intrusive enough to warrant radical behavior changes. Surveillance by the federal government, defending against terrorists and state enemies, and by law enforcement, in pursuit of criminals and evidence against them, is sometimes overly intrusive or abusive of human rights, but is largely supported by many citizens and 'is what it is.'

Third: Three entities are responsible for the current situation in which privacy is on the wane. Likewise, three entities can 'change the world': the government, commercial enterprises, and the people. It is neither possible nor preferable for any one of these three groups to solve the privacy problems raised by the current situation, as all three have played important roles in making it 'what it is.' If we are to 'change the world,' all three entities—the government, commercial enterprises, and the people—can and should take purposeful, meaningful, and prompt action to ameliorate the concerns that exist within the current situation.

Fourth: Our democracy is at risk, sooner rather than later, if we don't work together to enact solutions and return privacy to its rightful place as a basis for our democratic form of government.

This fourth stipulation requires explanation and defense. This claim hearkens back to Marshall McLuhan's treatment of information via his "medium is the message" metaphor. McLuhan's point was that although communication content is seemingly the focus of our interactions, communication content is less important, overall, than is the infrastructure that supports the exchange of meaningful messages. McLuhan proffered the image of media content as the raw meat that a burglar throws over the back fence to distract the guard dog. The more engaged people become in conversations about content, the less attention they pay to the effects of the infrastructure. The medium 'massages' everything; this is especially true for the dominant medium in a given culture. The Internet is currently the dominant medium and is deeply implicated in the degradation of privacy.

We could argue at some length about specific practices—the content in the media. For example, I will make recommendations about privacy policies and terms of service for, and ways that data get collected from, Internet sites. McLuhan would probably issue soft chuckles and some smiles. Focusing on the content within a medium causes one to take one's eye off the ball. What is more important are the practices that the infrastructure enables; the medium's affordances, so to speak. That personal data can be collected—that they are collected—often without the conscious permission from, knowledge of, or control rendered by the subject of the data is the point of importance.

McLuhan often said that the dominant medium changes our sense ratios. Critics then argued over whether or not humans were actually physiologically changed by exposure to this medium or that. Again, they missed the point. Taking our current situation as an example, although the majority of Americans say they care about privacy, they do little to protect it within the new media environment. In other words, they don't pay any attention to or notice the damage that is being done to them. Their sense of indignation has been muted by the everyday, taken-for-granted, ubiquitous nature of activities that are out of sight and out of mind (and out of control, especially the control of everyday citizens).

It is fairly easy to take the point of view that 'this is how the Internet works'—that 'it is what it is.' That is especially true when the outcomes appeal to our reward systems. Many of the activities that we carry out via new media technologies are either entertaining, fun, interestingly engaging, or beneficial to our sense of efficiency. In other words, we really like the stuff. Again, the dominant medium has altered our sense ratios, in this case dulling our critical, analytical processes in favor of frictionless fun and ease.

Here, however, is the problem: The media infrastructures used by the government, law enforcement, operators of the Internet, mobile phone providers, credit card issuers, loyalty card program sponsors, and IoT manufacturers and service providers are now fully in place, taken for granted, and effectively operational. What is the problem? 'It is what it is.'

And so it seems—as long as the control of our great country remains in the hands of supporters of democracy, at best, or benign dictators, at worst. Throughout our history, whether our leaders supported democracy or seemed like petty dictators was not much of an issue. If we discovered that the leader and his cabal were less than democratic, we could simply turn the scoundrels out. After all, if push comes to shove, and the ballot does not work, we can always have a second American Revolution.

But what happens if/when our leadership is truly incompetent, illegal, or evil *and* they are in control of a total surveillance infrastructure? What would have happened to America if Richard Nixon had fully understood the implications of—and had full control over—the electronic recording resources in the Oval Office? If he had been able to 'bug' the DNC without using incompetent 'plumbers'? If he had been able to turn total surveillance power, as we have in today's intelligence community, over to J. Edgar Hoover's FBI to use against civil rights leaders, anti-war activists, and the press? What would happen in America now if we elect a president who is mean, insulting, vindictive, and poorly prepared to serve as president? What would happen in America now if an unprepared, bellicose, self-centered candidate took the reins of control over our surveillance state?

Not long after the election, the new president issues a press release reminding citizens of the importance of this president's first State of the Union address. Regardless, you have a busy life and you are not much interested in politics, so you fail to watch the speech. You get up the next morning and prepare for your day, head for your car in the garage, turn the key—nothing happens. The car won't start. A text message flashes on the smart-dash awareness system:

> Fellow citizen, it appears that you failed to watch or listen to our President's very important speech last night. Will be happy to play the audio for you now or you can watch the video on your smart device. After you have completed your review of this important information, we will return control over your automobile to your able hands.

Our current media and communication infrastructures would require only modest modifications to enable this level of intrusion. Most of the cable systems that provide access to television and Internet can be switched to monitor our viewing habits if the companies want to make that change (or if the government forces them to). The IoT will include most of our devices, including our cars.

Of course, I'm fairly sure that we will never elect such a person. Perhaps democracy in the United States should worry more about a terrorist attack that triggers a 'step up' from the Patriot Act and the exponential expansion of the NSA (as we experienced after 9/11) than about the likelihood of electing a despot.

The point, of course, is that the *medium is the message*. We have enabled, across the board, an infrastructure that could/would support tyranny (if put to that use) to an extent never before possible in the United States. We have so obliterated limits to the collection, exchange, commodification, and dissemination of private information that we have participated in the construction of an infrastructure that can be used to negate democracy. If leaders—political, military, or commercial—were of a mind to do so and were situated properly in the political arena, the infrastructure could be turned against citizens in ways that would make targeted marketing and advertising look like Aesop's Fables or Mother Goose's nursery rhymes. Our democracy is at risk because we have lost track of the infrastructure of our dominant medium in favor of reveling in its content.

In the June 7, 2016, online edition of the *New York Times*, Quentin Hardy notes that Tim Berners-Lee wants to develop a new version of the World Wide Web because the current version, which he helped design and develop, "controls what people see, creates mechanisms for how people interact. ... It's been great, but spying, blocking sites, repurposing people's content, taking you to the wrong websites—that completely undermines the spirit of helping people create" (Hardy 2016). Berners-Lee and a number of digital pioneers of the early Web, including Vint Cerf and Brewster Kahle, led the Decentralized Web Summit on June 8 and 9, 2016, in San Francisco. The group responds to the crises they see surrounding the Internet, including the fact that "[t]he current Web is not private or censorship-free. ... The Decentralized Web aims to make the Web open, secure and free of censorship by distributing data, processing, and hosting across millions of computers around the world, with no centralized control" ("Call to Action" 2016). In other words, we do not have to accept that the current digital environment 'is what it is'; we can indeed 'change the world.'

I am mindful that some things cannot be changed; the stipulations at the start of this chapter probably go too far for some privacy advocates. Nevertheless, I do not much enjoy books that make unreasonable proposals in the face of straw-men-presented-as-bugaboos. I will try to constrain my recommendations to the doable, perhaps even the preferable, but always within range of the possible.

Given the content in this book, it is silly for any reader to expect me to recommend anything but the following: **Everyone should follow the five FIPs.** It is what it is. We can change the world. Unfortunately, both the recent past and the present make clear that this 'simple' solution will probably not be enacted.

Having gotten that out of the way we can look at a series of recommendations for ways to not only improve the situation in the short term but move strongly against an infrastructure that could be used to support the overthrow of our way of life.

As you will soon see, we *can* begin with the five FIPs. We can propose that they do serve as a reasonable template and set of goals. While rejecting the simple admonition that 'everyone should just follow them,' we cannot turn our backs on their essential reasonableness. My recommendations are divided into three sections, one for each of the entities: the government, commercial entities, and the people. Organizing them this way does not imply that the entities work separately; obviously, they are interdependent. Appropriate references to the FIPs will be made along the way.

Action Proposals for Government

In general, Tanner and others propose that "public records laws that enable data brokers to exist in an unregulated way need to be changed" (Tanner 2014: 267). Few privacy advocates are satisfied with a totally unregulated data marketplace. The recommendations that follow are based on the premise that modest amounts of regulation from the 'invisible hand of the government' could improve our privacy situation without presenting excessive burdens to commercial entities. Recommendations for executive leadership, federal agencies, congressional legislation, the intelligence and law enforcement communities, court systems, and state-level governments follow.

Executive Leadership

The Obama administration's efforts regarding personal privacy in the new media environment are very difficult to evaluate. Avoiding contentious and inappropriate political commentary here, it is not difficult to suggest that President Obama faced difficult challenges attempting to push legislation through the Republican-controlled Congress. From one perspective, the fact that the Obama administration proposed comprehensive privacy reform at all is remarkable. Noting that it fell short in many ways may seem unfair in light of the president's difficulties with a recalcitrant Congress.

The 2012 document *Consumer Data Privacy in a Networked World: A Framework for Protecting Privacy and Promoting Innovation in the Global Digital Economy* attempted to build on existing law to fill in gaps, supplementing existing frameworks and extending protections to segments not currently covered by federal statutes.

> The framework provides consumers who want to understand and control how personal data flows in the digital economy with better tools to do so. The proposal ensures that companies striving to meet consumers' expectations have more effective

ways of engaging consumers and policymakers. This will help them to determine which personal data practices consumers find unobjectionable and which ones they find invasive. (*Consumer Data* 2012: 7)

A major section of the proposals, the Consumer Privacy Bill of Rights, is based on an updated view of the original FIPs:

> The Consumer Privacy Bill of Rights advances these objectives by holding that consumers have a right to:
> - individual control
> - transparency
> - respect for context
> - security
> - access and accuracy
> - focused collection
> - accountability
>
> The Consumer Privacy Bill of Rights applies to commercial uses of personal data. This term refers to any data, including aggregations of data, which is linkable to a specific individual. (*Consumer Data* 2012: 10)

Implementation of the plan was to take place via a series of processes that the government would sponsor with multiple stakeholders:

> The Federal Government will work with stakeholders to establish operating procedures for an open, transparent process. Ultimately, however, the stakeholders themselves will control the process and its results. There is no Federal regulation at the end of the process, and codes will not bind any companies unless they choose to adopt them. The incentive for stakeholders to participate in this process is twofold. Companies will build consumer trust by engaging directly with consumers and other stakeholders during the process. Adopting a code of conduct that stakeholders develop through this process would further build consumer trust. Second, in any enforcement action based on conduct covered by a code, the FTC will consider a company's adherence to a code favorably. (*Consumer Data* 2012: 24)

The FTC was identified as the primary enforcement mechanism in that the process would lead the stakeholders through the development of codes of conduct and then task the FTC with enforcing against violations of those codes.

I recommend that persons interested in privacy within the US socioeconomic realm should read the White House white paper. The work is fairly thorough, relatively clear, apparently well-intentioned, and on-point overall. Noting its shortcomings is not intended as damning critique. As noted by Omer Tene, associate professor at the College of Management School of Law, Rishon Le

Zion, Israel (and affiliate scholar at Stanford University), "To be sure, if Congress heeded the White House's call for legislation codifying the FIPPs, US privacy law would be radically reformed." (Tene 2013: 1236; also see Halbert and Boyd 2012).

There are however, multiple problems with the approach to executive leadership demonstrated by the white paper. First, the report is limited to the commercial data marketplace. On the one hand, of course, this makes sense: Different sets of rules apply to our security situation than to our commercial enterprises. We must have spies, and they must do their business of protecting us from our enemies; we must have the economy, and it must do its business to produce innovation and profit in a capitalist economy. Obviously, the same rules cannot apply to both segments.

However, by totally excluding comments about the use of data in the security environment, the administration failed to recognize what had, by 2012, become an obvious fact (and would later become both an Achilles' heel and a searing reality): Government and commercial data streams cross, and any discussions of privacy and big data must include analysis and proposals for both sides of the equation (government/security and commercial/targeting). Leaving the government's role out, when it is so deeply entwined in the commercial collection of private data, damns the recommendations as something that 'they' (the commercial side) should do instead of facing the reality that the changes must be taken together. Further, such an approach elides the government's complicity and, therefore, seems disingenuous.

This difficulty would be made abundantly clear in 2014, when President Obama spoke about the mass surveillance crises brought about by the Snowden affair. There, "Mr. Obama spoke eloquently of the need to balance the nation's security with personal privacy and civil liberties," but he "did not address the bigger problem that the collection of all this data, no matter who ends up holding onto it, may not be making us any safer" (*NYT* Editorial Board 2014). In other words, not two full years after the recommendations of the "commercial-data-only" white paper, it became patently clear that the government's role in data collection is every bit as important to privacy as is that of the private sector.

Second, "the White House Blueprint's legacy is restricted to a sluggish multi-stakeholder self-regulatory process" (Tene 2013: 1237). The proposal reminded one of a planning document for a training session targeted at people who would lead training sessions for people who would lead groups. One can suspect that the administration knew that any substantive legislative proposals were dead on arrival at the steps of the recalcitrant Congress. Nevertheless, the articulation of a broad series of multi-stakeholder meetings that would lead to developments of codes that those stakeholders would then voluntarily implement and follow seemed, from the outset, to be a waste of time and energy. The idea was to lead a virtually infinite number of stakeholders through a process of coming to agreement on an iteration of the FIPs that they had ignored for over a decade. Updating and refining those principles was probably not going to be enough to convince stakeholders to immediately adopt them. In short, the white paper was destined to be treated as little more than political grandstanding from the very start.

Third, the focus of the report, and its primary recommendations, was on enabling consumers to 'better understand' what's being done to them, and with their data, rather than encouraging businesses to change their data collection practices. The status quo, default position, appeared to be "This is the way it is, so we should learn how to do it better." This is not a surprising position for the leader of one of the world's principal capitalist economies to take, especially at a time of continued economic pressures at home (and from abroad). However, starting the privacy conversation with the notion that companies should be allowed to collect virtually any data they want as long as they do so 'properly' more than likely fails to offer substantive suggestions as to how to curtail abuses by commercial operations. While it is true, and stipulated here, that the Internet is not going away, and that targeted marketing works to produce profit, risks to our democracy from mass surveillance, data collection, and the data marketplace justify approaches that seek to amend rather than assume 'business as usual.'

The recommendations for executive actions proposed here echo campaign promises made by then Senator Obama: (1) protect American security without violating citizens' privacy, and (2) reform commercial practices to protect citizens' privacy:

> The candidate Obama spoke clearly, directly strongly and without equivocation about protecting civil liberties and not giving up our freedoms. "I will provide our intelligence and law enforcement agencies the tools they need to take out the terrorists without undermining our Constitution and our freedoms. That means no more illegal wiretapping of American citizens. That means no more national security letters to spy on Americans who are not suspected of committing a crime. No more tracking citizens who do no more than protest a misguided war. No more ignoring the law when it is inconvenient. That is not who we are. That's not what is necessary to defeat the terrorists." (Manick 2013)

Strengthening privacy laws is precisely what Obama pledged during the 2008 presidential campaign. He told CNET at the time that: "I will work with leading legislators, privacy advocates, and business leaders to strengthen both voluntary and legally required privacy protections." Obama's privacy promise was a cornerstone of his technology platform. His campaign Web site pledged that as president, he would "strengthen privacy protections for the digital age." (McCullagh 2011)

Given that the risks to our democracy have increased exponentially during the eight years of the Obama administration, and due to the increasing ubiquity of surveillance, data collection, and the data marketplace during that time, future presidents should adopt the position that protecting Americans' privacy is a primary goal and responsibility of the executive. This responsibility exceeds the constraints of mere campaign promises.

Specifically, the president should revoke authority for mass surveillance on Americans that was authorized by presidential letters and issue no additional authorizations. In-place writ and warrant systems should be used to authorize data collection on Americans. Further, the president must lead efforts to rein in the data marketplace and commercial intrusions into privacy. Cheerleading, grandstanding, and campaign promises are inadequate executive actions in the face of these pressing threats to democracy. Recent administrations have virtually turned constitutional principles upside down in defense of the country against terrorist attacks. Our democracy is no less at risk from the anti-privacy infrastructures that we have created than it is from attacks by external forces. *The president should recognize the threat and meet it with both symbolic leadership and substantive executive actions.*

Federal Agencies

The FCC has begun to exert regulatory authority over aspects of the Internet. Enhancing privacy is one of the early targets of FCC attention. After establishing the Internet as a telecommunications service, the FCC moved to encourage net neutrality. Opponents, especially telecommunications companies, challenged those actions in court. The extent of the FCC's regulatory authority has yet to be fully worked out.

In proposing rules for third-party set-top box manufacturers, FCC chairman Tom Wheeler insists that new regulations would require that "even companies that have made behavioral data a big business would have to obey privacy rules that are similar to those that govern the cable industry today" (Fung, "Third-Party Cable" 2016). The new regulations would require third-party set-top box manufacturers to follow current rules that "prohibit cable companies from collecting 'personally identifiable information' on consumers without first getting their consent. They also prevent cable companies from sharing that data with third parties without subscribers' written or electronic consent, except when it is necessary for providing cable service" (Fung, "Third-Party Cable" 2016).

A number of observations apply in this regard. First, cable technologies are responsible for bringing television programming and Internet access into significant numbers of American homes. The data contained within set-top boxes are a virtual gold mine of information about private consumer behavior. Regulatory limitations on what cable companies can do with that data are an important protection. Second, one can note that cable companies are not damaged or threatened by the regulation. Extending the regulation to third-party set-top box manufacturers will not damage them either. The government's protection of consumers' rights does not hinder the ability of content and service providers to deliver products or services. The regulations simply restrict the companies from using consumer data in the data marketplace without consumer consent.

The FCC has shown that it can play a meaningful role in protecting consumer privacy across a broad range of new media enterprises. For example, the FCC sanctioned Verizon for using 'supercookies' without consumer permission. "As part of the settlement, Verizon Wireless was fined $1.35 million and is required to notify consumers of its data collection program, as well as get permission from users before sharing consumer data with third-party partners" (Kang, "Verizon Settles" 2016). Again, the regulatory action does not prevent the corporate use of private data; rather, it requires that the company notify the consumer and get permission for the data collection and data usage.

On March 31, 2016, the FCC voted to move forward with a proposal for new broadband privacy rules. "The proposed rules would require ISPs to obtain consumers' consent before drawing on their Web-surfing data for behavioral targeting. The basic framework of the proposed rules ... requires broadband providers to obtain consumers' opt-in consent before using data about them for [most] ad purposes" (Davis 2016). Broadband providers immediately slammed the rules as confusing for customers.

The range and scope of FCC regulation in new media spaces is yet to be determined. *That the FCC should play a role in protecting consumers in these activities is consistent with its historic oversight of radio, television, telecommunications, and the news business in American cultural history. The fact that the Internet* was excluded from FCC oversight is probably more of a historical glitch than having been the right thing to do. Further, the fact that so many entities in the data marketplace—data scorers, data merchants, and a range of other businesses—are unregulated highlights the role that the FCC can play within the media enterprises that provide private data to the marketplace.

The FTC can also play meaningful roles in the data marketplace, without depending on overwhelming amounts of regulation. The FTC can take action against firms (including regulated new media firms and unregulated companies in the data marketplace) that employ deceptive practices and other fraudulent activities. The FTC is particularly concerned about three aspects of data privacy: deceptive ToS, unfair privacy policies, and inadequate protection of, and security for, personal data that companies collect and trade.

The authority of the FTC to regulate these matters has been validated, repeatedly, across a number of cases and incidents. For example, in 2015 "a federal appeals court ruled that the Federal Trade Commission (FTC) has the power to take action ... against companies that employ poor IT security practices" (A. Richards 2015). That authority is found

> under 15 U.S.C. Sec.45, part of a 1914 law that gives the FTC the power to prohibit "unfair or deceptive acts or practices in or affecting commerce." FTC Chairwoman Edith Ramirez wrote, "Today's Third Circuit Court of Appeals decision reaffirms the FTC's authority to hold companies accountable for failing to safeguard consumer data. It is not only appropriate, but critical, that the FTC has the ability to take action on behalf of consumers when companies fail to take reasonable steps to secure sensitive consumer information." (A. Richards 2015)

The FTC envisions itself as one of the primary governmental enforcers of privacy regulations. Commission chairwoman Edith Ramirez insists "a company's failure to provide reasonable data protections constitutes an unfair practice, because we think it's a reasonable expectation for a consumer." Further, she notes that in regard to privacy policies and ToS, "We expect companies that make promises to actually fulfill those promises. If a company makes a particular promise in their privacy policy or through some other mechanism, we expect them to comply" (Peterson, "How the Country's" 2015).

Later recommendations in this volume target judges and the courts as the appropriate entities to make meaningful changes in the ways that the government interprets the relative fairness of privacy policies and, especially, ToS. As is noted, however, the status quo finds courts unwilling to employ reason in the face of unfair wrap contracts. *In the absence of rational court action, the*

FTC may be able to provide a secondary venue for relief from the deceptive practices that have become taken for granted in the wrap contracts of new media outfits.

In 2011, former FTC staffer Chris Soghoian wrote, "Most consumers do not read privacy policies, and often would not understand them if they did. In fact, the mere presence of a privacy policy is often misunderstood by consumers to mean their privacy is protected. ... While 'clickwrap' consent to a privacy policy may create a contract, it makes little sense to allow obscure boilerplate agreements to determine the existence of a reasonable expectation of privacy under the Fourth Amendment" (Davis, "Do Users Read" 2011). In 2014, the FTC took action against Atlanta-based PaymentsMD for collecting health information for uses other than those intended by patients when they signed consent forms. "According to the FTC's complaint, consumers consented to the collection of their health information, but further authorizations were buried in fine print in a small window. Consumers signed off by clicking a box in the window. The FTC alleged that wasn't sufficient disclosure, and consumers reasonably believed the authorizations were used only for billing purposes" (Dwoskin 2014).

The FTC has shown that it is able to enforce reasonable regulations supporting privacy while also remaining sensitive to businesses aspirations for innovation and economic viability. For example, in 2012 the FTC sanctioned MySpace for deceptive practices related to promises made in its privacy policy.

> Myspace's privacy policy promised that the site would not share users' personally identified information, or use information in a way that was inconsistent with the purpose for which it was submitted, without giving notice to users and receiving their permission. Myspace, however, provided advertisers with persistent unique identifiers, called Friend IDs, of users who were viewing particular pages on the site. ... The proposed settlement bars Myspace from future misrepresentations about users' privacy, requires the company to implement a comprehensive privacy program, and requires regular independent privacy assessments over the next 20 years. (Gross 2012)

In 2015, the commission was faced with requests by privacy advocates to sanction Uber for similar practices (Farivar, "FTC Asked" 2015). However, after carefully examining Uber's privacy policies and practices, the FTC communicated with the company to help shape its behavior rather than imposing immediate sanctions. In the process, the FTC demonstrated sensitivities to business interests while remaining vigilant concerning consumer privacy rights.

> In a speech at Fordham University Law School, FTC Chairwoman Edith Ramirez warned that imposing "legacy regulations on new business models" can stifle

competition and ultimately leaves consumers worse off. But, she said, regulators shouldn't shy away from enforcing important consumer protections on issues like health, safety, or privacy. "We must allow competition and innovation in the form of these new peer-to-peer business models to flourish." ... Any new regulations might not necessarily come from the FTC itself. While the commission does have authority over issues like privacy and data security, it also offers advice to state and local agencies on how to impose regulations without hurting competition. (Sasso 2015)

As noted, judges and courts have been unwilling to interpret overly complex and unfair wrap contracts as impediments to consumer rights. The FTC has not been as hesitant (or clueless) over the matter and has indicated a willingness to act on behalf of consumers in this regard. Privacy in the new media environment would be improved, by leaps and bounds, via meaningful changes to privacy policies and ToS. *The FTC's continued oversight and enforcement roles are strongly encouraged, particularly regarding fairness in privacy policies and ToS and strong stewardship when data are mismanaged by corporate entities. The FTC should continue encouraging enterprises that collect data to do so only after receiving permission from consumers who have been made aware of what data are collected and the uses to which the data will be put.*

Congressional Legislation

Congressional Legislative Action on Commercial Activities

The two laws that currently provide the structures that follow the FIPs, in business practices that are very much like the privacy marketplace and new media, are the FCRA (1970) and FACTA (2003). Numerous experts point to the viability of these laws and their usefulness in terms of protecting consumers.

It is useful to note that these regulations influence the ways that credit reporting agencies and other financial institutions treat private data. The regulations constrain the agencies from certain predatory practices as well as mandating specific consumer rights. While it seems somewhat self-evident at this point, with one law from 1970 and the other from 2002, we should highlight the fact that neither regulation put the credit reporting operations out of business. The regulations specify consumer protections and enable consumers to file for full credit reports from the three largest credit bureaus annually—at no charge to the consumer. And yet these businesses thrive; the invisible hand of the government has not significantly altered the market in negative ways by protecting consumers.

The FCRA "promotes the accuracy, fairness, and privacy of information in the files of consumer reporting agencies" ("A Summary of Your Rights" n.d.). Consonance with the FIPs is notable:

- You must be told if information in your file has been used against you.
- You have the right to know what is in your file.
- You have the right to ask for a credit score.
- You have the right to dispute incomplete or inaccurate information.
- Consumer reporting agencies must correct or delete inaccurate, incomplete, or unverifiable information.
- Consumer reporting agencies may not report outdated negative information.
- Access to your file is limited.
- You must give your consent for reports to be provided to employers.
- You may limit "prescreened" offers of credit and insurance you get based on information in your credit report.
- You may seek damages from violators.
- Identity theft victims and active duty military personnel have additional rights. ("A Summary of Your Rights" n.d.)

In amending the FCRA, FACTA articulates additional requirements for credit agencies and specifies additional consumer rights:

- identity theft prevention and credit history restoration
- protection and restoration of identity theft victim credit history
- improvements in use of and consumer access to credit information
- enhancing the accuracy of consumer report information
- limiting the use and sharing of medical information in the financial system
- financial literacy and education improvement
- protecting employee misconduct investigations. (FACTA 2003)

Some fundamental weirdness exists within this regulatory regime. Many of the companies that are constrained and encouraged by the FCRA and FACTA also participate in the broader data marketplace without regulation. In other words, even though it is clear that the constraints and encouragements provided by these consumer protection regulations do not damage the credit reporting industry, the regulations apply only to activities directly related to financial services like banking, insurance, and real estate; that is, the regulations apply to credit reports. The very same companies participate in the data marketplace and are not constrained, within the marketplace, with regard to all of the other personal data that they exchange.

The three credit reporting agencies required to provide free annual reports and constrained by federal regulations—Experian, TransUnion, and Equifax—are three of the largest players in the unregulated data marketplace industry ("Opt Out List" n.d.). The notion that the very same companies that are required to provide annual credit reports, at no cost, to consumers are free to generate enormous profits in the data marketplace using information drawn from the very same consumers, in a totally unregulated way, is ludicrous beyond logic. FTC member Julie Brill, among many others, has noted that one way to improve data privacy and consumer rights is to strengthen the FCRA and FACTA to include data brokers and the data marketplace. "Many tech firms are calling on the government to allow them to reveal how and how often the government seeks information about individuals. We ought to demand the same sort of transparency from the commercial data brokers that know much more about us than we do about them" (Brill 2013).

Congress can, and should, require major participants in the data marketplace to make reports about data collection and usage available to citizens, much in the way that the credit report provision is now mandated.

It is unreasonable to expect that American commercial new media companies will comply with European laws rather than the laws of the United States. It is equally unreasonable to assume that congressional legislation will tightly follow European law as a template. It is also unreasonable to assume that American socioeconomic and political philosophies will shift such that American legislation will protect everyday citizens' privacy rights more than corporations' rights to profits.

However, observing the outcomes of European approaches and legislation is informative and provides both models and test cases for needed adjustments to American regimes. For example, the right to be forgotten (discussed in chapter 4) has now been progressively in place in the EU for over two years. In that time, Google has variously resisted and cooperated with the protocols. The experience has made clear that Google is fully able to manage providing citizens with information about the data that it processes and that the protocols are not so overwhelmingly unfair as to force Google out of the EU marketplace. In fact, despite its various efforts to resist the protocols, Google has not only complied, but has expanded its cooperation almost to the extent that few would claim the law is ineffective.

Ordered by the EU's high court in 2014 to allow consumers access to information about data collection and information found as a result, as well as to enable consumers to correct inaccurate information (or information that is not in the public interest), after 22 months "Google has responded to European Union data watchdogs by expanding its right-to-be-forgotten rules to apply to its search websites across the globe. ... The multinational said that it would use geolocation signals (such as IP addresses) to restrict access to delisted URLs on all of Google's search domains," not merely those within EU-specific domains (Fiveash 2016).

No claim is made here that the EU rules are perfect or should be adopted in total by American lawmakers. However, the FIPs are simply inoperable without empowering consumers to discover what data has been collected and how they are being used. The European Union's "right to be forgotten" protocols are only a small piece of this puzzle. The removal procedures have been vigorously debated. On various grounds, Google complies with only about half of the removal requests that it receives. Freedom of speech and freedom of information advocates insist that erasures and removals might constitute efforts to manipulate social, cultural, and historical records.

Nevertheless, one of the troublesome features about the World Wide Web is that it has not learned how to forget. Computers and networks spread and archive information exponentially and ubiquitously. In the United States this feature is coupled with, and complicated by, the safe harbor provisions of the Copyright Act (DMCA) and of the Communications Decency Act (Section 230), which protect holders of private information/data (ISPs and Web services providers) from prosecution for violations committed by third parties. Removing information from databases and websites is very difficult—sometimes simply not possible, or extremely expensive and time consuming. Once accomplished, privacy violations are extraordinarily difficult to remediate.

Legislation that moves the American scene in the direction of the protocols modeled by European approaches would improve privacy protections for United States citizens. The European experience, with Google and the right to be forgotten, is clear evidence that American companies can do more, and can do better. Specific recommendations for their behavior follow in a later section of this chapter. American lawmakers, however, should more often approach these questions with the common understanding that 'it's a computer—it will do anything you program it to do.' If American lawmakers make laws that protect the privacy of American citizens, most American new media enterprises will find effective ways to implement the necessary changes without going out of business.

Congressional Legislative Action on Intelligence Activities

Developing a series of national privacy laws is probably too much to ask of Congress. In fact, it may well be that privacy advocates should be very careful about what they ask for, because they just might get it. Significant security legislation these days tends to look a lot like the Patriot Act. Nevertheless, one of the starting places for discussions about American privacy regimes is that there is precious little national law. In the cases of the privacy torts, few federal regulations harmonize the various state protocols. Some would argue that the American privacy scene will not be protective of its citizens until there is national legislation. I will not call for any on the commercial side of the equation. However, the security side demands more careful attention than Congress has so far engaged.

It is simply not possible to discuss this aspect without raising 'political' questions. I will take what I hope is a relatively straightforward position: I do not believe that we can stay on a war footing forever against external terrorist threats. The "war on terror" is an interesting way of branding the political philosophy; however, it does not accurately describe the way Americans are or should live indefinitely. It is simply not possible to defend a country as large as the United States from every single potential terrorist threat. Citizens want safety and security by reducing risks and making sure that America is not an easy target for the many enemies who would do us harm. Those aspirations should not, however, move us to suspend constitutional protections for the American way of life—to essentially turn our backs on American democracy—under a slogan proven to be a convenient excuse for excessive executive and intelligence service power.

The courts will continue to variously rule; the intelligence committees will get briefings that either fully inform or partially obfuscate. What is needed is broader and deeper congressional interest and involvement in privacy and security with relation to executive power and a war footing. Executive and judicial branch orders for mass surveillance on Americans via secret-court-authorized and presidential-letter-mandated spying are unacceptable trade-offs between security and privacy. *Congress should act to require specific writs, either subpoenas or warrants, for surveillance on Americans.* We cannot remain at war, suspending basic rights, forever.

Senator Ron Wyden argues that expanded active spying hurts the American economy, that Congress should shut down government-ordered backdoors into digital systems, that affirmative action should be taken to ensure the security of the data of millions of Americans, and that the data Americans give to companies should be held and used by those firms rather than turned over to the government (or sold to other companies) (Farivar, "Senator" 2016).

Unfortunately, Wyden is one of very few congressional voices supporting the privacy rights of American citizens. *Elected officials should respond favorably to protecting American democracy by protecting citizens' privacy. Elected officials must guard against government-sponsored intrusions that lead toward governmental leveraging of new media infrastructures that could be used to establish totalitarian controls. Careful reexamination of the Patriot Act with a focus on protecting Americans' privacy as a primary way to defend democracy would be a good place to start. Adding oversight to the FISC/FISA court procedures such that the court's activities would change from 'secret' to 'closely held' would be a sound second step.*

Intelligence Community and Law Enforcement

It is especially difficult to make recommendations for policy and procedure changes that should be taken by the intelligence community and law enforcement. Regardless of whether recommendations target governmental, commercial, or personal entities, skeptics proclaim, "You know that *they* won't do that"—despite efforts to make those recommendations reasonable. Such skepticism is particularly strong with regard to the intelligence community and law enforcement. Recent history suggests that spies and the police will leverage every technology they can acquire (only sometimes lawfully) and use them in pursuit of criminals and terrorists, actual or potential.

To some degree, US citizens expect the intelligence community and law enforcement to act against people and organizations that seek to harm Americans. Technologies can help the intelligence community and law enforcement improve their abilities to prevent crimes and apprehend criminals. The intelligence community and law enforcement officials invoke legalities that demarcate the contours for allowable and acceptable protocols and most always claim that their activities are within the bounds of the law. They are particularly concerned that the methods they use produce materials that are acceptable as evidence in court. Law-abiding citizens would like to support the enforcement organizations that protect lawful ways of life.

History is replete with examples of the dynamics among the law, the courts, and intelligence communities and law enforcement operatives; intelligence and police work have long been done in ways that present repeated test cases of legal permissions and constraints. This new media age is no different in that regard. However, the advanced abilities of new media technologies—the vast capacities for data collection, storage, analysis, and sharing—raise the stakes in these negotiations of citizens' rights, democratic principles, and duties to protect and defend. Unfortunately, numerous recent incidents cast doubt on claims by the intelligence community and law enforcement about their dedication and adherence to legal structures. In order to 'keep up with and stay ahead of the bad guys,' our 'protectors' often push the boundaries of legal and acceptable practices. Too often, they go well over the line—and stay there, until the courts or legislatures push them back (sometimes after citizen protests). This section is *not* about corrupt spies or cops. Pure corruption among our 'protectors' is assumed to be less prevalent than the bulk of good work that they do. At issue here is their tendencies toward seeking expanded technological abilities for surveillance, data collection, and big data analysis.

There can be little doubt that the intelligence community and law enforcement officials wish to move forward using every available effective technology. It's also the case that many leaders and officials in these domains seek to do their jobs under the law, within the constraints provided by legislation and the courts. It has become difficult, however, to place full trust in the intelligence community and law enforcement with regard to their use of new media technologies and big data. There are simply too many instances when our 'protectors' seek to stretch constraints beyond the intent or spirit of the laws, to seek authorizations that leverage loopholes or look for special (often secret) consent from courts or officials, or to secretly use technologies that would not pass muster with the public or the courts or legislatures if the full extent of their existence and use was known and understood.

Potential solutions for these shortcomings are usually presented as recommendations for stronger laws and increased oversight. However, such calls have often been made and are just as often ignored. Further, operatives who protect us (law enforcement officials and intelligence agents) chafe at and resist unwieldy protocols that make their jobs more difficult. While privacy advocates and civil libertarians often make strong appeals for placing additional constraints on the intelligence community and law enforcement, many other citizens, especially political leaders, insist that our systems are not yet tough enough on criminals, or terrorists, or immigrants, or people otherwise different from the supposed American norm. More than a decade ago, the 9/11 attacks raised the stakes for 'keeping Americans safe,' shifting both the legal frameworks and the dominant cultural narrative away from protecting civil liberties and toward figuring out how to stop the bad guys before they can damage life, limb, or property.

Rather than make recommendations for sweeping changes in legal protocols, I will state some preferences for the ways that our 'protectors' should act. I connect these preferences to the major contention of this book: that business as usual, leveraging the powers of surveillance and big data analysis *against* American citizens and putting into place an infrastructure that enables ubiquitous privacy intrusions, threatens the American democratic way of life. Intelligence operatives and law enforcement officials are called to defend our laws, our Constitution, our democracy. When they act in contravention of civil rights, including privacy rights, they undermine the system that they have sworn to protect.

I recommend that members of the intelligence community and law enforcement should only collect targeted information about specific subjects in ways that are clearly authorized by existing law. Mass data dragnets, with the intention of looking for patterns or identifying movements that were previously not recorded, expose innocent citizens to unnecessary scrutiny. Our protectors should focus on credible threats. They should stop using mass data collection that compromises the privacy rights of innocent people. They should stop looking for loopholes and special set-asides to enable them to skirt the law.

One dramatic yet informative example should suffice. The Patriot Act was instituted in response to the 9/11 terrorist attacks. Additionally, the NSA received permission for massive buildups in data collection and analysis, resulting in the NSA collecting untold amounts of communication data on everyday Americans. Arguments over the secret presidential letters and secret court orders that enabled and authorized "The Program" are replete with political and spy craft intrigue. The arguments are ongoing, as are the surveillance programs. The funding mechanisms for the buildup and buildout are astronomical, especially in light of the fact that very few members of Congress are even vaguely briefed about the programs.

The various investigations into lapses leading to 9/11 have made fairly clear that collecting the phone conversations, email traffic, and Web-browsing habits of all law-abiding Americans prior to the attacks would have done little to stop the attacks. Rather, a number of practices internal to intelligence and law enforcement agencies and the Federal Aviation Administration (FAA) could have alerted authorities to the potential for trouble or would have found law enforcement/intelligence responding better or more strongly to the warnings and information they did receive.

FBI investigations after the attacks found that "the 19 suspected terrorists received flight training from at least 10 US flight schools. At least 44 people sought by the FBI for questioning received some flight instruction" (Fainaru and Grimaldi 2001).

Records indicate that FBI agents had interviewed operators of the flight training schools prior to September 2001, in many cases while in possession of the names of, and details about, foreign-born students. Subsequent investigations made clear the large number of linkages between the hijackers and Al Qaeda as well as other connections with Osama bin Laden and terrorist training camps and activities abroad. Many officials testified that there were good reasons why the FBI had been unable to connect the dots and prevent the attacks. No claim to the contrary is made here.

The point here is that the warning signs had nothing to do with the mobile phone, email, and Web-browsing habits of law-abiding American citizens. Collecting that information, as the subsequent and current NSA mass data collection programs accomplish, would not have provided solutions to the problems at hand in the late 1990s and early 2000s that contributed directly to the 9/11 terrorist attacks. Spending significantly more money on immigration procedures, tracking nonresident aliens, increasing intelligence abroad, tracking financial distributions from countries with terrorist links—these are activities that might have prevented 9/11 and that might enable intelligence services and law enforcement to thwart future terrorist plots. These aspects remind us of what Postman meant when he warned that we should carefully interrogate new technologies to understand what problems they are targeted to solve, whose problems we have identified, what new problems the solutions themselves cause, and who is advantaged and disadvantaged from the changes wrought by the so-called solutions. The NSA's "Program" does not solve a problem that the American people have (no strong evidence suggests that such a program either would have prevented the 9/11 attacks or will prevent future such events); causes many additional problems (abrogating the privacy rights of law-abiding Americans and undermining their trust in both the government and the commercial communication companies forced to cooperate); advantages intelligence and defense contractors and employees; and furthers the political agendas of people who support using the NSA to spy on Americans rather than constraining it to investigating foreign nationals.

Intelligence operatives and law enforcement officials should turn their attention to solving specific problems with technological solutions that target particular weaknesses or threats. This is good police work. The fact that a new tool is available does not mean that it is the best tool or that it has to be used. Sometimes, other considerations must come to the fore. In this case, protection of Americans' democratic liberty should win out over narrow threats.

As was pointed out in the section above, we cannot forever maintain a war footing against terrorism; doing so is both ineffective and abusive of democracy. Leaders of all stripes, including those of intelligence and law enforcement operations, should realize that many of the decisions made under stress after 9/11 were influenced by the devastation and its consequent fears. Bad decisions, policies, and laws resulted.

Our legislators have (largely) failed to avert their eyes from the fog of war and make better decisions about data collection and analysis practices that compromise our liberty. Leaders of our law enforcement and intelligence agencies now (often) carry out the results of those bad decisions. They should reexamine their reasoning and refashion their practices appropriately.

Most of all, our intelligence and law enforcement practitioners should adopt an age-old formula for decision making: The ends do not justify the means. Our 'protectors' are not keeping America safer by going about the business of destroying democracy in what is ostensibly the process of protecting us.

Judges and Courts

Legislative gridlock has gripped Washington, DC, and Congress for a number of years and for a wide variety of reasons. President Obama was effectively unable to provide decisive leadership against intrusions on privacy because of the war on terror and the political gridlock that made it difficult for him to propose anything other than collaborative actions between the government and commercial entities. Federal government has tasked two agencies, the FCC and the FTC, with primary responsibility for consumer protections in this area. In short, we are in desperate need of higher levels of carefully thought out judicial actions in support of privacy in the new media environment.

Unfortunately, our court system has mostly failed in this regard, and miserably. Not uncommonly, one side or the other on an issue expresses displeasure with court decisions that prefer the positions of the other side. My displeasure with the performance of the judiciary in matters of protecting privacy rights of citizens is shared by many privacy advocates and civil rights supporters. However, I fully recognize that these are positional viewpoints that include political inclinations and personal judgments about the law and policies, procedures, and practices; these inclinations are not shared by those on the other side of the issues nor by many judges who made decisions against these principles. Obviously, I'm going to suggest that the judges are wrong, shortsighted, and not supportive of democratic principles. These claims are intended neither as personal attacks against individual judges nor as a general lack of respect for the court or its systems. I do, however, join with other critics and analysts who believe that our judges have fallen far short of our expectations on a number of counts.

Contemporary judges face a number of difficult conundrums in the area of intelligence gathering. Secretive plans and actions by the executive and the intelligence community place many judges in no-win situations. The executive and the intelligence community cloak their actions within excessively expensive efforts to legitimize spying on Americans. In some instances, our judges are faced with actions bearing the executive stamp of approval or approval by the secret FISC/FISA court. Judges who believe in applying, rather than interpreting, laws have difficulty opposing executive orders; often, the intelligence community forcefully requests those orders. Gratefully, other judges have actively sought to interpret constitutional protections when considering cases with privacy and constitutional implications.

It would be easy to point out cases when judges decided against the government and in favor of American's privacy rights and also to recommend that more judges follow those examples (Hsu 2016). That is what advocates for privacy are supposed to do. These matters will be battled out in both the legislature and the courts for decades to come. Rather than chastise particular judges and celebrate singular cases (many of which are later reversed on appeal), *I recommend that judges increase their sensitivity to constitutional protections of privacy*. Again, our most sound footing is democracy rather than wartime extensions and exclusions.

I believe that our judges and courts are on much less sound ground in the commercial area than they are when attempting to apply laws and executive orders in matters of security. It appears that, on the whole, our judiciary has fallen far short of expectations with regard to adjusting to the new media environment. Some of this is to be expected—the legislative process has been so slow that courts are forced to apply old media law to new media circumstances.

However, it appears that judges and courts have exacerbated the situation by not moving quickly enough to recognize differences between the old and new environments. It appears that judges and courts have been unwilling to adjust to the new environment. In fact, it often appears that judges and courts have been unable or unwilling to fully inform themselves about the complexities involved in the new environment. The shortcomings are most pronounced in the areas of privacy.

I recommend that our judges and courts hold law enforcement to strict interpretations of constitutionally based privacy rights, rather than repeatedly interpreting new media activities as exempt from protection. An example is the unwillingness of judges to apply the general principles that protect wired telephone conversations to mobile phone devices and communication. I'm fully aware of the inability to shield wireless signals from interception. I'm also fully aware that people use their cell phones in public places and thereby demonstrate little or no expectation of privacy in those instances. However, to conclude that the ease of intercepting wireless signals and the carelessness of some users constitute proof that no warrants are required to intercept private phone conversations (and/or collect data about them) and to use that data for prosecutions and in the data marketplace is a misapplication of old principles to new media circumstances.

It takes a long time to legislate in the face of rapidly changing new media technologies. In the meantime, judges need to do more than just specifically apply existing laws as though they are direct analogs that can work in new media circumstances.

As noted earlier in this volume, *judges should stop treating the Internet and new media companies as nascent industries in need of protection and coddling.* Many rulings put concerns over stifling innovation ahead of commonsense observations that new media compromise citizens' privacy and democracy. Remediating one particularly glaring judicial weakness could bring about significant improvements across a wide swath of commercial privacy issues.

Judges must take a more informed and critical approach to terms of service and privacy policies. Online privacy policies and ToS cannot be treated using offline contracts as direct analogs. I am convinced that if judges hearing cases that turn on privacy policies and ToS forced themselves to do one simple procedure, they would change the ways they evaluate and rule on cases and thereby force commercial industry operatives to modify policies and terms. *Every judge involved in such cases (this would not hurt any judge, as a matter of preparation for cases they might hear) should spend one eight-hour day carefully examining the privacy policies and ToS of every single online and offline new media product and service they encounter.* Hit a Web page, read the documents; start your smart car, find and read all the fine print; have a smart electric/gas meter, read the disclosures; watch TV or get Internet from a cable service, read the policies; go online and use LexisNexis for case analysis/documents, read its policies. I'm pretty convinced that, like everyone else in our culture, judges don't engage even a fraction of the privacy policies and ToS that they are responsible for during their everyday activities. And in not doing so, they usually rule that the particular set they have examined, over the weeks- or months-long course of a case, is reasonable and legally binding. If consumers only had to deal with one set at a time, they might be able to navigate the policies and terms and make good decisions. But consumers do not encounter policies and terms, or the products and services to which those policies and terms apply, one at a time. Further, having a long look at all of the policies and terms might convince judges that it is totally unreasonable to equate wrap contracts with old-fashioned paper contracts. Doing so abuses citizen rights, consumer protections, *and* common sense. In this regard, our judiciary share primary responsibility for stripping (particularly) online privacy from American citizens via the shoddy ways they have sided with corporate articulations of one-sided privacy policies and ToS.

Kim notes that wrap contracts fail in at least the following three ways: First, they lack reasonableness. "Courts expect too much from consumers, and far too little from companies that draft these agreements" (2013: 181). *Judges must refocus their approach as to what "reasonable" means with regard to wrap contracts. Courts currently interpret it to mean that it's possible for customers to read ToS and privacy policies (that is, 'they can if they want to').* The fact that no one reads them (and few could understand them even if they did) indicates the current judicial interpretation of reasonable to be wrong. *Instead courts should force wrap contracts to be presented in a way that one can reasonably assume that they will be read and understood* (Kim 2013: 176). Second, wrap contracts fail to tailor assent. Wrap contracts should not present an all-or-nothing proposition, yet courts are "generally reluctant to evaluate the fairness of a bargain or the adequacy of the consideration" (Kim 2013: 192). *Courts should allow for tailored consent.* Wrap contracts should provide a middle ground, between 100 percent agreement and none. That bifurcated choice forces consumers to either agree or not use the site/product/service, and the World Wide Web is not built to encourage users to just go away. Third, wrap contracts should impose only those restrictions that are required by the transaction, rather than a wide range of obligations upon or entitlements from the consumer. Kim notes that "the form of the contract reflects (or should reflect) its function. In other words, intended user function of the contract should be determined, or at least signaled, by its form" (2013: 200). *Courts should limit wrap contract enforcement to restrictions that are required by the transaction.*

Online participants don't read privacy policies and ToS; most would not understand privacy policies and ToS if they read them. Mere notification that privacy policies and ToS exist and can be found by "clicking here" does not equal agreement. Merely using a site or service does not constitute agreement, especially when the only options for not agreeing are (1) trying to come to grips with overly long and complicated privacy policies and ToS that lead to agreement out of frustration and (2) being blocked from the site for not agreeing. Privacy policies and ToS do not spell out consumers' privacy rights, largely because the exceptions to them (e.g., "We will only share your data with our partners") are written more to conceal than to reveal commercial intent.

In short, our judiciary has totally failed in its duty to apply justice in cases involving consumer rights, including privacy rights, in commercial cases hinging on privacy policies and ToS.

In many instances, laws that applied to old media cannot be applied to new media with fidelity. Rather than stretch old laws beyond their credible scope, judges should make rulings based on constitutional values and good judgment. Judges should not attempt to write new law. However, judges entering rulings that will be appealed because the reasoning used is not fully based on existing laws force appellate courts, and eventually legislatures, to confront the inadequacies in decaying legal structures. Our laws never keep up with technological developments and the courts lag last. With the rapid changes in our new dominant mediums, processes for vetting legal structures and strictures must be accelerated. These reexaminations are started when judges turn against existing laws that lack justice or reason. Wrap contracts, which are even worse than that, are at the heart of how and why online gathering of personal information has gotten so very far out of line.

State-Level Actions

The lack of federal standards is one of the shortcomings in privacy law. Many pundits and critics of our current privacy environment note that not having federal legislation about privacy issues disadvantages all parties in these complex situations. However, federal privacy controls have obvious risks. Two come to mind immediately: first, the current legislative environment is so polarized and contentious that it is difficult to craft sound national legislation, especially about topics as controversial and sensitive as privacy; second, the federal government's involvement in massive surveillance programs targeting domestic American citizens raises significant doubts in the wisdom of allowing the federal government to propose and apply a broad range of privacy law. The bifurcation of the Obama administration's approach is illustrative in this regard: While willing to take meaningful, if not conclusive, steps encouraging commercial privacy protections, the administration was totally unwilling to participate in disciplining its own house (the federally sponsored surveillance state). Many would also argue that the administration's approach was overly cozy with commercial interests. These factors leave the states as the primary legislative protectors of personal privacy.

Actions taken by the states present something of a paradox in the area of privacy. As noted above, states are able to play a role that seems pragmatically out of reach to the federal government. However, local and state law enforcement operations depend on federal funding for much of their resource base. Federal entities provide funds for programs and equipment with the expectation that the local and state operations will provide data, information, and usage experience in return. Many local and state law enforcement agencies depend on federal funding; some of that federal funding is influenced by intelligence and military lobbying. Rejecting the material and political aid provided by federal intelligence and law enforcement agencies would put state and local law enforcement officials in a bind. While state and local lawmakers might want to limit and curtail surveillance, data collection, and data analytic law enforcement strategies, state-level law enforcement officials often operate with the opposite point of view and practices ("Privacy Legislation" n.d.).

I recommend three principles for state actions: (1) state governments can and should adopt approaches that are guided by the FIPs; (2) state governments can and should constrain government agencies and law enforcement from predatory practices; and (3) state governments can and should enact state-based protocols and procedures that increase consumers' rights in the commercial environment.

States and the FIPs: Constraint of Government Agencies and Law Enforcement from Predatory Practices

Among the many ways that states can both legislate effective privacy protection for their citizens and inform other states and the federal government as to appropriate actions and privacy approaches, the states can adhere to the FIPs. The state of Virginia provides an excellent example in its Government Data Collection and Dissemination Practices Act (GDCDPA). The GDCDPA opens by noting the importance of privacy and the growth and ubiquity of data collection and processing:

1. An individual's privacy is directly affected by the extensive collection, maintenance, use and dissemination of personal information;
2. The increasing use of computers and sophisticated information technology has greatly magnified the harm that can occur from these practices;
3. An individual's opportunities to secure employment, insurance, credit, and his right to due process, and other legal protections are endangered by the misuse of certain of these personal information systems; and
4. In order to preserve the rights guaranteed a citizen in a free society, legislation is necessary to establish procedures to govern information systems containing records on individuals.

The document then articulates principles for state agencies and functions:

C. Recordkeeping agencies of the Commonwealth and political subdivisions shall adhere to the following principles of information practice to ensure safeguards for personal privacy:
1. There shall be no personal information system whose existence is secret.
2. Information shall not be collected unless the need for it has been clearly established in advance.
3. Information shall be appropriate and relevant to the purpose for which it has been collected.
4. Information shall not be obtained by fraudulent or unfair means.
5. Information shall not be used unless it is accurate and current.
6. There shall be a prescribed procedure for an individual to learn the purpose for which information has been recorded and particulars about its use and dissemination. (GDCDPA 2015)

In 2016, state senator J. Chapman Petersen exemplified contemporary state-level legislative additions to Virginia's FIP-based statues when he proposed (still-pending) legislation that protects citizens against intrusive data collection by law enforcement:

SB 236. Government Data Collection and Dissemination Practices Act; Collection and Use of Personal Information.

Provides that, unless a criminal or administrative warrant has been issued, law-enforcement and regulatory agencies shall not use surveillance technology to collect or maintain personal information where such data is of unknown relevance and is not intended for prompt evaluation and potential use regarding suspected criminal activity or terrorism by any individual or organization. The bill authorizes law-enforcement agencies to collect information from license plate readers, provided that such information is held for no more than seven days and is not subject to any outside inquiries or internal usage, except in the investigation of a crime or a missing persons report. After seven days, such collected information must be purged from the system unless it is being utilized in an ongoing investigation. The bill also adds to the definition of "personal information," for the purposes of government data collection and dissemination practices, vehicle license plate numbers and information that affords a basis for inferring an individual's presence at any place. (SB 236 2016)

States can and should take action against a number of information-gathering activities that prey on citizens' privacy. Some states have proposed legislation to limit the kinds of information that federal intelligence agencies, like the NSA, can collect from state entities, by setting limits on the sorts of information that state law enforcement officials can collect. For example:

A new Oregon law prohibits state and local law enforcement officers from using "forensic imaging" to obtain information contained in a portable electronic device except with a warrant, or by consent. A similar law in New Hampshire prohibits law enforcement from obtaining location data from electronic devices without a warrant. And California Gov. Jerry Brown signed a sweeping privacy bill into law requiring a court ordered warrant or subpoena before law enforcement agencies can obtain or access electronic communication information or electronic device information from individuals or third-party providers. These states followed the lead of Utah after Gov. Gary Herbert signed HB0128 in 2014. The law prohibits Utah law enforcement from obtaining phone location data without a warrant and makes any electronic data obtained by law enforcement without a warrant inadmissible in a criminal proceeding. ("Privacy Legislation" n.d.)

A number of states have passed legislation limiting particular devices and thereby constraining the kind of information that can be collected (and passed on to other agencies, including the federal operations). As previously noted, a number of states have taken action against so-called 'stingray' devices. In similar fashion, a number of states have passed legislation limiting the use of license plate readers. One can expect a plethora of state-level legislative action about the use of drones. Currently, the most widely applied restrictions come from the FAA and relate to safety for airlines and airports. Local and state officials have expressed significant concerns about the finer-grained privacy aspects inherent in drone use.

Privacy advocates have long lamented the roles that states play in providing data to the commercial data marketplace as a way to increase local or state revenue. Again, the local and state entities are presented with a paradox: Cash-strapped communities and states see data as a potential revenue stream with an almost unlimited number of willing buyers; citizens seldom respond favorably to learning that their local and state governments have sold their data to the commercial marketplace. Again, since no federal standards exist, and each state sets its own policies and procedures, state-level approaches to open records laws vary wildly.

The National Association of Counties' *Open Records Laws: A State by State Report* (Winkler 2010) makes a thorough review available to the public. One can examine each state's policies and laws concerning open records, including information about how to request records and which kinds of records are available (or not) for commercial use. As previously noted, many states have instituted policies that limit the commercial use of data derived from public databases. However, many states still sell some information directly into the data marketplace and others limit usage only by way of charging commercial outfits for provision of the data.

It is difficult to know whether taxpayers derive significant benefits as a result of states receiving additional revenues from the sale of personal data. It is not difficult to notice how often citizens are upset by the news that their local and/or state governments are selling their personal data into the data marketplace with neither adequate notification nor their permission to do so. A general observation is that it appears that citizens don't like the practice and that, essentially, citizens want their local and state governments to use private data only for the purposes for which it was provided rather than as a revenue stream.

State-Based Actions in the Commercial Privacy Environment

On the one hand, the California law (the Online Privacy Protection Act of 2003 – California Business and Professions Code sections 22575–22579) leaves a lot to be desired, as all it requires of commercial data gatherers is that they notify customers of their data collection policies and practices.

On the other hand, California is a leader in developing and articulating forward-looking policies and procedures for advancing better privacy policy and law. The January 2013 document *Privacy on the Go: Recommendations for the Mobile Ecosystem* presents a series of principles adopted by leading mobile communications providers. The document articulates sound principles as direction for developers of mobile applications. *I recommend that additional states emulate these practices.* The recommendations include the following:

For App Developers
- Start with a data checklist to review the personally identifiable data your app could collect and use it to make decisions on your privacy practices.

- Avoid or limit collecting personally identifiable data not needed for your app's basic functionality.
- Develop a privacy policy that is clear, accurate, and conspicuously accessible to users and potential users.
- Use enhanced measures—"special notices" or the combination of a short privacy statement and privacy controls—to draw users' attention to data practices that may be unexpected and to enable them to make meaningful choices.

For App Platform Providers
- Make app privacy policies accessible from the app platform so that they may be reviewed before a user downloads an app.
- Use the platform to educate users on mobile privacy.

For Mobile Ad Networks
- Avoid using out-of-app ads that are delivered by modifying browser settings or placing icons on the mobile desktop.
- Have a privacy policy and provide it to the app developers who will enable the delivery of targeted ads through your network.
- Move away from the use of interchangeable device-specific identifiers and transition to app-specific or temporary device identifiers.

For Operating System Developers
- Develop global privacy settings that allow users to control the data and device features accessible to apps.

For Mobile Carriers
- Leverage your ongoing relationship with mobile customers to educate them on mobile privacy and particularly on children's privacy. (*Privacy on the Go* 2013: 2)

Proponents and opponents alike can find aspects of this document sufficient or insufficient. Nevertheless, it demonstrates that state-level governmental entities can work with industry representatives within their states and have a broad commercial impact. It is certainly more likely that state operatives (working with organizations and employers within their geographic area) can craft meaningful protections of privacy more quickly and more efficiently than can the federal government (by organizing massive numbers of meetings for constituency subgroups on the way to federal legislation that is probably dead on arrival).

This proposal from California is not intended as a complete version of all privacy legislation related to mobile devices. It does, however, suggest a state-based approach to protecting consumer privacy that can have a significant and lasting impact. However, neither federal nor state legislation can get out ahead of innovation in the development of technologies. Laws will always lag. Political controversies over social, economic, and political costs and benefits will always interrupt the development of meaningful legislation. Sometimes, though, the resulting legislation evokes the old saw "Be careful what you wish for, because you just might get it." The DMCA comes to mind as an example of a poorly constructed federal protocol that barely works in the new media environment that it was supposed to protect.

Commercial entities—the new media industries, especially their major innovators and leaders—will either change their direction with regard to protecting consumer privacy or will continue undermining democracy as we know it. Industries can be nudged with effective legislation; laws cannot turn back the new media revolution. Commercial entities have to take dramatic action if we are to recover privacy as a foundational value of American society. Leaders of commercial operations have claimed for years that their business models depend on the trust that American citizens bestow on their companies. In spite of that rhetorical flourish, few have acted in ways that give the companies a legitimate claim to that trust. Commercial entities will have to do better if they want to operate under democratic capitalism in the future. In many ways, our future is in their hands.

Action Proposals for Commercial Entities

Industry Leadership

There is no doubt that new media and telecommunications operatives have provided personal data to American intelligence and law enforcement agencies and operatives, sometimes willingly and, at other times, without full cooperation. Materials from the Snowden revelations validated reporting about specific incidents and ongoing practices and policies (see, for example, Brodkin 2013; Goodin, "Exposed" 2014; Savage, "CIA Is Said" 2013, "NSA Said to Search" 2013).

Apple's opposition to FBI requests to unlock the cell phones of the San Bernardino terrorists was not the first time a new media company had resisted data requests by the intelligence and law enforcement communities. There are numerous examples of such hesitancy on the part of leading figures in the new media industries. Particularly in light of the Snowden revelations, American technology companies are concerned that American consumers will associate commercial data collection with activities by the intelligence and law enforcement communities and thereby be less forgiving of commercial data collection. American technology companies have proffered a particular set of meanings for the phrase 'Americans need to trust us with their data.' In this regard, American tech companies want to be able to tell American citizens that they will not betray citizens' trust by turning over data to law enforcement or intelligence communities without being compelled by the law or court order. Unfortunately, it does not exactly mean 'You can trust that we will not misuse your data in the commercial marketplace.'

Twitter has long been a voice against excessive cooperation with the government and its intrusive data collection procedures. Along with other tech firms, Twitter prefers to resist the ubiquitous data requests from US intelligence and law enforcement. A particular point of contention is the government's limitations on tech companies' abilities to publicly report information about the requests (how many requests, for what kind of data, etc.). The lack of transparency forced on tech companies lowers their 'trust quotient' among the citizenry. Twitter sued the federal government in 2014, claiming that the gag orders violate their First Amendment rights against prior restraint of their free speech (Koh 2014). The case remains in litigation and may well make its way to the Supreme Court (Gershman, "Department of Justice" 2016).

Intelligence and law enforcement agencies prefer that US tech companies cooperate with government surveillance and data collection; the government does not like end-to-end encryption or not being given the keys to the backdoor for software, systems, and networks ("Director Discusses" 2015).

However, the tech community's newfound 'religion' on privacy is not without precedent. In 2014, Microsoft—a company that has cooperated with many government data requests—challenged a search warrant issued by a US federal judge that would compel the company to hand over customer email data stored on its servers in Ireland. Microsoft countered with the claim that email data contained on the company's servers, located in Dublin, are subject to Irish rather than US law and that those servers should thus be out of reach of US intelligence agencies, law enforcement agencies, and courts (Silver 2014). Then in April 2016, Microsoft upped the ante exponentially by suing the US Department of Justice over its frequent use of secrecy letters that bar companies from disclosing to customers the fact that (in this case) Microsoft has been served with, and has responded to, warrants permitting the government to read customer email. In a lawsuit filed in the federal court in Seattle, Microsoft's home jurisdiction,

> the company asserts that the gag order statute in the Electronic Communications Privacy Act of 1986—as employed today by federal prosecutors and the courts—is unconstitutional. The statute, according to Microsoft, violates the Fourth Amendment right of its customers to know if the government searches or seizes their property, and it breaches the company's First Amendment right to speak to its customers. Microsoft's

suit, unlike Apple's fight with the Federal Bureau of Investigation over access to a locked iPhone, is not attached to a single case. Instead, it is intended to challenge the legal process regarding secrecy orders. (Lohr 2016)

In March 2016, prosecutors with the US Department of Justice (in a case apparently not related to terrorism) took on WhatsApp, the messaging app recently purchased by Facebook. The *New York Times* reported, "In the last year, the company has been adding encryption to those conversations, making it impossible for the Justice Department to read or eavesdrop, even with a judge's wiretap order. As recently as this past week, officials said, the Justice Department was discussing how to proceed in a continuing criminal investigation in which a federal judge had approved a wiretap, but investigators were stymied by WhatsApp's encryption" (Apuzzo, "WhatsApp" 2016).

In light of the Snowden revelations, and with the San Bernardino incident still close in the rearview mirror, Apple has begun tasking its engineers and developers to implement server systems, internal to the company, that could be protected against various types of government intrusions. The company is not only concerned about data requests; it is concerned over evidence that intelligence agencies have intercepted hardware and software prior to installation in order to modify configurations, making systems easier for the government to hack (Brodkin 2016).

Innovators and developers bringing the Internet of Things/Everything (IoT) to fruition have particularly important roles to play in determining the future state of privacy and democracy in America. *Companies with particularly large stakes in the IoT—IBM, Cisco, and Google, for example—should develop and sell the IoT on a privacy rather than a sharing model. Companies should work to make their systems the most secure and private operations available. IoT firms should seek to win the market race by protecting our privacy while still enabling the IoT to deliver promised and potential advantages.*

Implementing the IoT within the current infrastructure, considering its weak protections of personal privacy, is dangerous business. Already, at a juncture prior to intensive implementation of IoT technologies, invasive and downright creepy practices illustrate the dangerous path we are on, as infrastructure for surveillance and control is being installed under the guise of advertising and marketing.

> Drivers along a busy Chicago-area tollway may have recently noticed a large digital billboard that seems to be talking directly to them. It is. Launched last month here as well as in Dallas and New Jersey, the eerily Orwellian outdoor campaign for Chevy

Malibu uses vehicle recognition technology to identify competing midsize sedans and instantly display ads aimed at their drivers. ... "This is just the tipping point of the disruption in out-of-home," said Helma Larkin, CEO of Posterscope, an out-of-home communications agency that designed the Malibu campaign with billboard company Lamar Advertising. (Channick 2016)

Executive Leadership

Apple CEO Tim Cook's position against the FBI's request drew a public battle line that had previously been articulated in (mostly) closed-door negotiations between industry representatives and government operatives over the years. Cook claimed that the government's wanting Apple to undo its own encryption was "bad for America" (Frankel 2016). Numerous tech companies defended Apple's position in the case (see "Amicus Briefs" n.d.).

As noted, the two sides have long clashed over encryption standards. In the main, technology companies have only cooperated with privacy intrusions by government agencies when forced to do so (although some of the complicity—such as AT&T's—that was uncovered by the Snowden revelations brings the supposed unwillingness of that cooperation into some doubt). Since the Snowden revelations, tech companies have been less willing to cooperate overtly and have been more publicly outspoken about the differences between their approaches to privacy protection and the government's.

Unfortunately, American tech companies have a serious credibility problem in this regard. It is exceedingly hypocritical for American tech companies to claim to be the protectors of consumer privacy. In fact, such claims are ludicrous. Tech companies are more interested in protecting the appearance of integrity than acting with it. American tech companies collect and leverage as much or more private data then do the law enforcement and intelligence communities combined. And they do it for profit rather than with a motive to capture bad guys or prevent terrorist attacks. Still, the newfound religion—commercial firms supporting privacy—is to be commended, with hopes that it continues and expands.

Two of the biggest and most influential forces in this regard are Google and Facebook. Their founders still play important roles in the day-to-day operation of their businesses. In both cases, early manifestations of the software and platforms operated without advertising revenues (Auletta 2009; Kirkpatrick 2010). In fact, both sets of founding entrepreneurs actively opposed and rejected the integration of advertising with their products and services. At a point in their developmental history, when rising operating costs and investors' RoI needs required higher levels of profit than could be produced without advertising revenues, both companies evolved into advertising-based operations. As Web 2.0 (and beyond) added ever-increasing amounts of data to the mix, both companies flourished by leveraging as much data as they could collect, such that each now collects more data than most other operations on the planet.

Google and Facebook, and their founders, are fabulously rich. Neither company, nor their founders, would suffer from returning to their original roots. Both companies could survive and flourish even if they developed and dedicated shards of their operations that did not collect massive amounts of data from and about their users. I'm well aware that leveraging data is central to their DNA as well as to their current business plans. However, both companies have evolved well past needing to operate in the usual ways.

The founders of Google (Mr. Brin and Mr. Page) and of Facebook (Mr. Zuckerberg) should consider dedicating alternative shards of their services and products that provide similar customer outcomes as are derived from their advertising-based operations (search and social connectivity), without using customer data for anything other than ensuring excellent search results and empowering social connectivity. Both of these companies could provide their featured product—search functionalities in the case of Google and personal networking functionalities in the case of Facebook—to millions of users without moving the data taken from users into the data marketplace. Facebook and Google could provide their core functionalities via systems separate from their primary products, thereby offering consumers alternatives to data mining for advertising and marketing. The market could be allowed to decide which products and services consumers prefer and the founders would be able to return to their original entrepreneurial visions. Leadership by these two new media behemoths would go a long way toward showing other technology outfits that it is possible to provide digital products and services without abusing customer privacy. If new media companies really want to get the government off the backs of the people (and keep them out of our private lives while protecting our democracy), they must reexamine their contributions to an infrastructure that is too often used against citizens.

As noted earlier in this chapter, Tim Berners-Lee, who developed HTML and who is credited as the inventor of the Internet, strongly believes that action by tech leaders can and should make "efforts at creating greater amounts of privacy and accountability, by adding more encryption to various parts of the web and archiving all versions of a web page. ... [This] would make it harder to censor content" (Hardy 2016).

Following the adage 'think globally, act locally,' we must note that it is not enough for local business owners and operators to wait for online behemoths to change the 'industry' privacy protocols before adjusting local behaviors. Because local business operators can make an enormous impact on our privacy environment in many ways, they must send a message for change to the new media industries. *The owner/operator of an enterprise is ultimately responsible for their organization's orientation toward privacy and the ways that data provided by the public are treated. The buck stops with the local owner/operator.*

A series of challenges to owners and operators of local businesses follows: Does your organization follow the FIPs? Or does your organization only pretend to protect privacy by hiding behind the legal language of your ToS—language that really only protects you and your organization and doesn't follow the FIPs? Do your practices offer your clients/customers the free chance to opt into the various ways that you use their data? Or must clients/customers take specific actions to opt out, without the real ability to do so (beyond not using your product/service because opting out disqualifies them in one way or the other)?

Is your excuse for not following the FIPs based on socially responsible reasoning? Or does it just come down to "We didn't know about them" (well, now you do) or "We need to make more money and it's too darned hard to do all that stuff and nobody really cares anyway"? It is certainly the case that not following the FIPs is easier and more advantageous to the bottom line than following them is. But since we consumers are mostly unable to protect ourselves, distrustful of the government's overprotection, and have failed to be protected by the bottom-line-oriented market, we need local community business/enterprise leaders to feel and exercise the social responsibility to protect our personal privacy.

In the long run, one wonders if local enterprises benefit from taking advantage of their customers or from not caring about customers enough to follow the FIPs and protect the personal information of their most trusted clients from abuse and misuse?

Let's imagine the following scenario. You own or operate a business. A customer comes through the front door. Before providing your product/service, you ask them to stop and to empty the contents of their pockets onto a table. Once they've done so, you take them into another room and provide your product/service. But while you are both gone, a bunch of folks (not sure who or how many) go through the contents on the table and grab what they want. After your transaction, the customer returns to the table, collecting their belongings on their way out the door.

One additional thing: The folks who raided the table left a small 'tip' for you in an unmarked envelope. Your customer didn't see that, but you know that every transaction will include both the outsiders and the tip.

A business owner/operator might read this scenario and say, "Absurd. I wouldn't do that to my clients/customers." But of course, when a business directs customers/clients to "like us on Facebook" or "follow us on Twitter," that's exactly what is happening: Local customers expose their personal contents (information) to a vast data marketplace, most likely without their conscious awareness and control, at the behest, and for the benefit, of a local business. The business gets 'free' advertising, public relations, customer engagement, and maybe even a revenue stream from the social networks and data marketplace; the customer/client gets ... what? You might tell them they get an "optimized Web (or shopping or cultural) experience." But what they really get is their pockets picked by data collection practices that they do not understand and cannot control. And this scenario doesn't even begin to address the nuts and bolts of what local businesses do with the information collected from clients/customers/users during transactions or how they calculate the value of further leveraging that personal data within the data marketplace (beyond the social network).

The FIPs are not complex or difficult to understand. They do not represent government intrusion or the collapse of the open/free marketplace. They stand as a symbol of our mutual respect for one another's fundamental rights to control our personal and private information. *Local business owners/operators are the first line of defense against inappropriate commercial intrusion into our private data and lives. They should work hard to protect their customers' privacy.*

Worker Behaviors

Although technical innovation is making solid progress toward artificial intelligence and self-replicating computing systems, at this point in the evolution of new media, people still make new media. Sometimes it is easy to say "the industry should do this or that," when what we really mean is "the people working in the industry should (or should not) do this or that."

Just as we teach students about privacy and the dire situation facing our democracy due to intrusive data collection, analysis, and implementation, it is useful to remember that we do so as a way to get them to act differently than they otherwise might once they enter the job market. It makes sense, then, to comment on actions that particular employees already working in the new media industries should take or not take. In other words, often the recommendations are for actions not that the industry should take, or that a company should take, but that the people working in technology industries should take. Historically, innovators in computers and networks were hackers and sometimes anti-establishmentarian. We may have reached the point when individual new media workers need to examine their motivations and the orders they receive from employers in light of the outcomes of intrusive data practices.

Recall that long before NSA contractor Edward Snowden orchestrated the distribution of massive amounts of classified documentation verifying and extending reports of agency wrongdoing and massive data collection, Thomas Drake, William Binney, J. Kirk Wiebe, and other NSA employees made numerous attempts to alert their supervisors to the constitutionally questionable practices. When repeatedly thwarted, they became unwilling whistleblowers, at enormous personal costs (Eisler and Page). Dramatically, in 2006, Mark Klein, a retired AT&T communications technician, corroborated claims that AT&T had "violated federal and state laws by surreptitiously allowing the government to monitor phone and internet communications of AT&T customers without warrants" by knowingly allowing the NSA to attach technology to switching boxes that routed copies of domestic phone traffic to the NSA (Singel 2006).

When faced with the possibility of being required to unlock iPhones in the San Bernardino case, a number of Apple employees were apparently willing to oppose government demands—perhaps even if those demands were eventually accepted by Apple—that they cooperate and unlock the targeted iPhone. Before the FBI employed hackers to unlock the phone while the agency was still pressuring Apple do so, "Apple employees [were] already discussing what they will do if ordered to help law enforcement authorities. Some say they may balk at the work, while others may even quit their high-paying jobs rather than undermine the security of the software they have already created, according to more than a half-dozen current and former Apple employees. Among those interviewed were Apple engineers who are involved in the development of mobile products and security, as well as former security engineers and executives" (Markoff, Benner, and Chen 2016).

The reasons why workers execute rather than question or resist corporate directives are almost infinite; most relate directly to getting and maintaining employment. Further, many workers in new media fields do not agree, in fact or philosophy, with privacy advocates. Many strongly believe that privacy is over-sold, or that the benefits of big data analysis and personalization far outweigh the negative costs. My effort to convince current and future workers otherwise intends no offense to their judgment or good intentions. *I recommend, however, that tech workers carefully examine the materials presented in this book, and from other sources, and thoughtfully analyze their own work. This is not a good time to disconnect the analytical logic we use in business from our values, morals, and ethics. Further, it's a great time to push ourselves to examine, more carefully than usual, the second-order effects and unanticipated consequences of constructing and implementing an infrastructure for ubiquitous electronic surveillance.*

Protocol Changes

The recommendations that follow can be mandated and implemented, variously, by agreements among high-tech corporations (such as through trade association standards), mandated by executives of one or more companies, or carried out by line-level workers. The recommendations reflect changes to standard protocols that either directly compromise the privacy of consumer data or contribute mightily to the degradation of consumers' abilities to protect themselves by controlling their private data in new media environments. Presentation order does not indicate hierarchy.

Privacy Policies and Terms of Service

As noted, wrap contracts in new media environments are incredibly predatory on private citizens/customers. Corporate lawyers write privacy policies and ToS for the exclusive purpose of protecting the company, often at the cost of obfuscating, eliding, or outright denying proper levels of consumer protections and rights. This practice has to stop. This practice puts corporate entities at the throats of their customers. Consumer protection only works when businesses and consumers work hand in hand. Consumers ought not defraud companies; companies should able to protect themselves from consumer fraud. However, the status quo of wrap contracts so favors corporate interests as to abrogate the rights of consumers, particularly in the area of privacy.

Technology companies are going to have to take action to reverse this trend; the courts have proved unwilling or unable to protect citizens in this regard. It is in the interests of businesses to begin protecting citizens' privacy. As we've seen in the preceding section, businesses believe that they need their customers' trust. At some point, technology businesses must walk the walk in addition to talking the talk.

Lawyers who write wrap contracts should be tasked to write them such that the contract protects both of the parties in the transaction. Wrap contracts should not be written to be purposefully confusing; they should not be displayed in ways that are purposefully difficult to find; and their language should be clear to lay audiences. Notification that privacy policies and ToS exist does not constitute agreement on the part of consumers who see those notices. When the consumer doesn't read the policies, they cannot rightfully agree to them (regardless of whether they've checked a box that says "I agree"). As long as wrap contracts are overly long, difficult to find, and hard to understand, consumers will not read them but will continue checking the "I agree" box. These actions merely mean that the consumer wants to get on with whatever it is they want to do (complete a transaction, see content on a Web page, sign up for a service). This does not mean that the consumer agrees to the terms (regardless of how contemporary courts have misconstrued those actions).

The lawyers and designers who work for new media companies are highly proficient at writing unfair wrap contracts. They should put an equal (or greater) effort into making sure that those policies are fair and easy to understand so that consumer agreement (checking the box) actually relates positively to understanding and willing agreement.

Flip the Defaults

Numerous academic studies and treatises have repeatedly indicated the importance of default values. Defaults exert powerful forces that nudge consumer choices and behaviors (Kahneman 2011; Thaler and Sunstein 2009). New media companies have repeatedly shown that they prefer to take advantage of customers by presetting defaults in a position favorable to the company rather than allowing or enabling consumers to set defaults based on decision making.

Presetting defaults to the advantage of companies is a predatory behavior that should stop immediately. When presets are leveraged in this fashion, companies show themselves to be taking advantage of their customers purposefully, self-consciously, and in ways that almost certainly disadvantage the consumer in favor of the business model of the corporation.

It is exceedingly easy to leave defaults up to the customer rather than presetting them to the company's preference. Further, actions on the part of new media companies to preset defaults to invade privacy and collect data without permission are unethical. Companies cannot establish trust with their consumers by taking advantage of them at the start and then hoping they will not discover and change the default.

New media companies should offer clear designations for options and make explicit the choices that consumers have for setting defaults. Presets could begin as opt-out, such that consumer rights are protected initially; corporate communication and persuasion can then be applied to convince consumers to change default settings to opt into practices, protocols, or services. Alternatively, settings can begin in a neutral position, with an instruction that consumers need to choose to opt in or opt out. Defaults always preset to have consumers opting into privacy-intrusive activities, such as data collection or sharing, are ethically indefensible.

Offer Multiple Versions

This recommendation was articulated in the "Executive Leadership" section above. One additional thought is added here. New media firms, especially online enterprises, have long worked hard to make their products and services accessible. Even foundational Web design courses feature discussions about accessibility; advanced studies and production courses enable students with the skills to bring a vast array accessibility features to websites and Web-based products. In the early days of the Internet, many website operators offered multiple versions of their sites, generally working toward compatibility with the capabilities of the machines and networks consumers used to access the Web pages. Occasionally one still sees the vestiges of this practice, particularly when multiple versions of sites are offered to customers in high versus low bandwidth or mobile versus fixed network situations.

Increasingly, however, contemporary companies are called to a higher level of effort with regard to site accessibility. For example, in March of 2016, a blind man in California "won a disabled-rights case against a Colorado-based luggage retailer accused of failing to make its commercial website accessible to the visually impaired" (Gershman, "Court Orders" 2016).

In similar fashion, Web designers and services providers could be tasked to present multiple versions of a site, with one version featuring a 'privacy-protection mode.' As is always the case, 'the devil is in the details,' and in this instance, commercial operations do not have a good track record. Wide-ranging efforts at "do not track" have been met with opposition by most online entities. The voluntary process from which companies can simply absolve themselves weakens "do not track" almost to the point of absurdity. Likewise, industry leaders work feverishly to develop bugs and supercookies, of all sorts, to defeat users' efforts to block tracking (hoping in vain to protect their privacy). If people in new media industries are going to work directly against customers and against consumers' wishes, the resulting situation is not going to breed trust by optimizing protections for consumer rights or democracy. *Voluntary provision of privacy-protected versions might not be a realistic goal, though it is certainly a recommended approach.* The FTC may need to step in and mandate such accommodations, much as courts require accessibility. Voluntary cooperation by industry operatives would be preferable.

Reconsider Data as Revenue Streams

According to Tanner, "the best businesses give consumers a choice whether or not to share their data, and offer benefits in return" (2014: 171). The Caesar's Palace and Harrah's companies operate tremendously successful casino operations in Las Vegas and elsewhere. Tanner documents their decisions, running from the 1990s through at least 2006, against collecting outside data—information that they could have combined with the mountains of information they have about their customers (2014: 40). Since 2006, Harrah's has supplemented its data with third-party information and has seen its bottom line improve. However, the company is careful about the types of third-party data that it uses and makes every attempt to narrow its data collection toward its primary business goal: improving services for specific customers who have shown that they want to deal with the casinos (Tanner 2014, 171–180, 209–223).

New media companies should collect data from their particular customers in order to provide products and services to them. Earning revenue by selling data into the data marketplace should be taken out of the business plans/models of these companies. Businesses should make money from their products or services rather than by selling customers' data into the marketplace. In the infrequent instances when a firm intends to sell customer data into the marketplace, the enterprise should seek explicit permission from an informed consumer. In that case, the operation may have to offer the customer a premium, discount, or other reward in order to obtain the customer's permission to use their private data for profit in the commercial marketplace.

Reengineer Data Anonymity

As noted, one of the primary 'lies' told by new media industry operatives is that, after collection, personal data are aggregated such that consumers' identities are anonymized. The truth of the matter, of course, is that anonymization is no longer in place, because various operators in the data marketplace provide de-anonymization services that reconnect people to their data. Additionally, of course, people assist this effort by using devices that help connect the dots—for example, smartphones accessing the World Wide Web, membership cards connecting retail purchases to credit cards, and GPS-enabled devices connecting phone calls and purchases to location data.

Companies developing and implementing ways to re-anonymize data would be a great aid to privacy and commerce alike. Many problematic privacy issues become less severe when the data do not identify specific individuals. Although most advertisers and marketers would resist, seeing re-anonymization as a move backward (to the days before Web 2.0), doing so would actually be a technological move forward, as the status quo blocks users from taking control of their own data. Providers of new media products and services want consumers to use networks with trust. Secure data environments will enable sellers to sell and users to find them. That's what the Internet is for. Re-anonymizing data will enable particular buyers and sellers to do business without unwanted intrusions by competitors and the government.

Nonprofits: Great Responsibilities

The recommendations that I will make for nonprofit organizations cause me emotional distress. These recommendations are difficult to make, as nonprofits find themselves in challenging social and economic situations. My basic premise is this: Nonprofit organizations, schools (especially institutions of higher education), social agencies, hospitals, and industry-supporting institutions of all sorts (such as supporters of the arts, education, health care) almost always lack adequate funding and thus depend on contacting large numbers of supporters and participants to raise their organizational profiles and increase their resource base. In some cases, nonprofits also utilize interactive new media technologies to keep in touch with their students, clients, customers, and patients. These technologies sometimes help nonprofit organizations serve their constituencies.

Nonprofit organizations, just like their for-profit counterparts, find themselves immersed in socioeconomic circumstances where 'everyone' is using a wide variety of new media technologies, especially Facebook, Twitter, Instagram, Pinterest, and other social media platforms and software. As an example, it is often easier for small nonprofit organizations to maintain a Facebook page than it is for them to develop and maintain their own website. It is far easier for nonprofit organizations to contact a wide population of their constituents by issuing messages to their Facebook or Twitter followers than it is to design and execute a communication campaign that effectively reaches those people. Time and efficiency are money, and money is not something that nonprofits have to waste.

In some cases, nonprofit organizations might even derive financial benefits from connecting their constituents to social media or other professionalized electronic platforms.

Regardless, I propose that in most instances, nonprofit organizations do not gain enough value from sending their constituents to privacy-invasive enterprises to offset the privacy risks their customers are thereby exposed to. Further, I propose that many nonprofit organizations are supposed to be the *most* protective stewards of our privacy; in many cases—education and health care, for example—nonprofit entities go to great lengths to protect the privacy of their constituents within their own organizations. However, many, if not most, of our most trusted nonprofit organizations compromise their constituents' privacy in a wide variety of ways, by sending them to privacy-invasive social media platforms.

The recommendation here is that nonprofit organizations, and especially their research divisions, ethics supervisors, and leaders, should do a very thorough and informed privacy audit focused on the ways that they use social media and electronic communication resources. In many cases, such an analysis will show that an organization going to great lengths to protect the privacy of its constituents, and to follow the laws in this regard, undermines those efforts by exposing those same constituents to privacy invasions that compromise (sometimes even reveal) the material the organization has worked so hard to protect. An example is appropriate.

Recall the privacy policy and ToS instance that I raised in chapter 3 concerning UnityPoint Health–Methodist. After filing a complaint with the health network about ambiguities in its privacy policies and ToS, and engaging in two phone conversations with various operatives in its customer service and HIPAA compliance offices, I received a letter from the local UnityPoint–Methodist HIPAA Privacy and Information Security/Compliance Officer. In that "CONFIDENTIAL" letter, the officer assured me that the company had completed a legal review of my concerns and that

> The law requires we would not receive remuneration for disclosing protected health information without the patient's authorization, so it was generally included in our consent form. However, we have confirmed with the regional marketing director for UnityPoint Health that we do *not* disclose or sell our patient information to a third party. Furthermore, the Patient Authorization for Release of Information will

be changed system-wide and the new form will not include language about remuneration. The change in the consent is being implemented at the corporate level and is currently in various stages of implementation. (Moore 2015; used with permission)

A number of lessons can be taken from this incident. First, even large nonprofit operations—UnityPoint–Methodist is a multi-state, multi-hospital, multi-provider network—can be responsive to consumer feedback, including complaints about both the letter and spirit of privacy policies and ToS. While there is no guarantee that organizations will be responsive, they may well be, so consumers should read the documents (fine print and all) and make sure that they understand and agree with what they are signing. Their signatures indicate understanding and consent, especially when these documents are presented on paper.

Second, UnityPoint–Methodist is a good example of a nonprofit organization with a strong preference for protecting the privacy of its customers—in this case, patients.

Third, UnityPoint–Methodist should consider continuing its practice of presenting privacy and ToS agreements on paper. While many network services providers (doctors and clinics, for example) still provide paper copies, the main branch of UnityPoint–Methodist Hospital in Peoria, Illinois, now 'goes through' the policies electronically at check-in. A customer service representative asks if the patient knows about and understands the hospital's privacy policies and, after the patient agrees, the representative presents a small electronic device and stylus for signature. The patient does not see or read the agreement they are signing, yet their signature indicates agreement (and stands as legal assent). This practice is not in keeping with otherwise strong privacy protections. It is fast and convenient for both parties, but it gives in to the notion that patients don't read the terms anyway instead of providing proper information that allows patients to give informed consent.

Having established UnityPoint–Methodist as a nonprofit provider that cares deeply about the privacy of their patients' data, I want to make a third point illustrating a conundrum for all nonprofit organizations. UnityPoint–Methodist has established a dedicated records and communication system that manages communication among patients and medical providers, including communicating with doctors and accessing financial and medical records. MyUnityPoint has been established as a secure means of communication; in doing so, UnityPoint–Methodist acquires a degree of control over its communication situation, theoretically enabling the organization to use protocols that protect patients, partner providers, and data.

However, the contemporary nature of the interactive new media environment—focused as it is on mobile smart devices—creates difficulties for even careful and caring data stewards. Patients who set up a MyUnityPoint account can connect that account to their smart devices. Accessing MyUnityPoint from a computer, let's say from the home or office, would connect only an IP address with inquiries into the system. However, accessing the MyUnityPoint system from mobile devices opens those exchanges to monitoring by a broad range of entities far out of the reach of UnityPoint–Methodist's good intentions and protective protocols. Patients use those devices habitually.

When consumers use mobile fitness-monitoring apps and services to manage their health, exercise, and diet activities, the data they send to the app developer's cloud are not protected by HIPAA or any other privacy law. Likewise, the mobile connections made to MyUnityPoint are unprotected. Access via mobile phones enables the collection of location data, identity data, timestamps, and so on, by entities outside the exchange between UnityPoint–Methodist and the patient. Data re-identifiers can use that information to connect patients' healthcare interactions with other data in their digital dossier. For example, people who access their health records may then go to a pharmacy, or call their health insurance company, or go on the Web to order healthcare products. When the MyUnityPoint system sends an automated message to patients each time their medical records are updated (patients can enable this function in their preferences), the system that UnityPoint–Methodist has installed in order to help protect patient privacy works specifically against that goal.

I am not proposing that nonprofit organizations must remove themselves from participation in electronic communications and records (that ship has sailed) or that they should avoid sharing patient data with the people it comes from (patients) in accord with the FIPs. I am suggesting, however, that the new media infrastructure is complex enough to motivate even the most careful organization in the nonprofit sector to take extra caution when engaging their clients in new media interactions.

In that regard, it is very difficult for me to issue a different recommendation to nonprofits than I will to private individuals: If you want to protect privacy and if you are serious about protecting our democratic system, do not use social media. When medical patients, counseling clients, social services clients, arts organization members, current, past, or perspective students at institutions of higher education—when any of these people are encouraged by their organizations to "like us on Facebook" or "follow us on Twitter," the privacy of those clients/

customers/patients/students is put at risk and nonprofit organizations vote in favor of the data marketplace, big data, and the surveillance infrastructure.

In some ways, I find it remarkable that organizations that are supposed to be at the forefront of progressive societal movements—and that should be the most informed about the dangers in using advanced technologies—are instead making decisions based on thinking that because their customers use social media, they have to be there too. I am baffled as to why institutions of higher education, which should be teaching critical approaches to media use, are instead climbing on the social media bandwagon while it tumbles over the cliff into the abyss of a surveillance state.

Some of the smallest nonprofit entities are at the mercy of young new-media technologists who do not think deeply about these issues. Many small nonprofit organizations can barely afford to pay for their Web services, let alone pay for a widely experienced and deep-thinking Web services manager. And of course, engagement in the new media environment is not the first or only example of organizations buying into technology just because they can or because it has become 'the way we do things.' I'm reminded how often my colleagues and I subject our students to death by PowerPoint.

In many ways, we depend on entities in our nonprofit sector to behave better than their counterparts in the commercial, for-profit segments. Taking the profit motive out of products and services should lessen, somewhat, the urge to sell out or give in to the culture. However, nonprofits have to stay in business, too; nonprofits cannot turn their backs on making a living, paying the bills, staying above water, making money. The new media environment is filled with opportunities for organizations to increase their bottom line and decrease their costs. Organizations that cannot work nimbly in the new media environment may not be able to survive.

Unfortunately, none of these economic realities change the privacy equation very much. Nor do they change my recommendations. *Nonprofit entities should work extra hard to avoid exposing their clients, customers, and patients to the ubiquitous privacy risks presented by the data marketplace.* Since I strongly believe in the old adage "physician, heal thyself," I should mention my business: higher education, specifically the departments I am associated with, Interactive Media and Communication Studies. Our students are preparing to enter professional careers in these fields; we are training them to do jobs that will involve them in the collection of user and customer/audience member data.

We have a responsibility to teach students as much about protecting privacy as we teach them about compromising it, no matter how much the collection and exchange of private data is "just the way the industry does things." We have a responsibility to do better than sending students into the data marketplace unaware of and unprepared for how to improve the privacy environments in which they will live. More specifically and emphatically, our students will be responsible for making the environment into what it becomes; our students will help the government and industries collect, analyze, and use private information. The question is not whether private information will be collected; information is the lifeblood of new media. The question is whether the collection, analysis, and implementations of that data are in accord with the FIPs.

Institutions of higher education also have a responsibility to follow the FIPs themselves. Just as with healthcare providers' adherence to legal and ethical constraints, higher education organizations work assiduously to protect students' private data. Institutions of higher education follow the mandates of the Family Educational Rights and Privacy Act (FERPA) as well as HIPAA (when dealing with students' medical records). And yet, colleges and universities tend to be much less vigilant over digital practices that compromise student privacy; for example, these institutions recruit via social media, use off-the-shelf commercial software in class applications, require students to interact with one another via social media platforms, and require that students, staff, and faculty leave cookies enabled on their browsers in order to access university-sponsored websites (knowing fully that most people are unlikely to change those settings for other Web work).

In another way, I think that institutions of higher education, or at least one segment therein, present an excellent example of the actions needed to protect privacy in this extraordinary age. Academic libraries are faced with a complicated and unprecedented privacy challenge. Libraries keep track of information—it's what they do. Libraries had to evolve faster than most other segments of higher education as they came to grips with the explosion of information and informational sources resulting from computers, computer networks, and the Internet. Libraries updated their systems, retrained their people, refocused their missions, and faced an enormous amount of change, adjusting more quickly and more efficiently than most entities within higher education.

And then, the roof fell in. As a result of the 9/11 terrorist attacks and the Patriot Act, libraries faced an existential conundrum. With the mere presentation of a particular kind of letter, libraries are required to turn over records of their customers' reading habits to law enforcement or intelligence community operatives. Libraries and librarians would have to do so without informing the customer—not then, not ever—that the request had been made. They would not be able to tell the customer what information was turned over or to whom.

They would not be able to challenge the request, either in person or in court. In short, libraries were to act as though they are instruments of the government with regard to the reading habits of customers who have (for whatever reason) come to the attention of law enforcement or the intelligence community.

What to do? How does one protect customers from such invasive practices? The solution that academic libraries came up with was dramatic and counterintuitive, running somewhat against their DNA. Many libraries no longer keep track of the checkout records of their customers once a given loaned/borrowed item has been returned. Of course, each library has electronic resources for keeping track of checkout behavior. But if they kept track, they would have to make those records available on request and could thereby compromise constitutionally protected privacy rights that have been abrogated by the Patriot Act. Instead, they chose not to keep the records after items are returned. As a result, they can refuse to provide information simply because they do not have it. Even better than Apple saying, "No, we will not help you unlock that iPhone by writing new programming," local librarians whose libraries have implemented this protocol can say, "No, we don't have that information; we don't store or retain it."

How much easier it would have been for librarians to say, "You know, there won't be many of these requests; we may not get a single one. And you know, we are in the information business and we've always kept track of checkout data and we are now better able than ever to do that because we have computers. This is the way the world is these days; of course we will provide those records when asked. It is what it is."

And yet, as true patriots and defenders of democracy in the face of the new media onslaught against privacy rights, they did not give in to the culture of surveillance. This is a model for and a lesson to other nonprofit agencies. They may have to take extraordinary actions in defense of privacy and our democracy; these actions might even contradict their fundraising or other organizational goals. *Our nonprofits have to lead in the culture, not merely follow the commercial impulse. They must run the privacy calculation on behalf of their patrons always leaning toward protecting rather than compromising privacy rights.*

WE, the People

As noted throughout, people—everyday citizens—alone are not able to solve the privacy problem in the new media environment. Users are only one portion of the three-sided problem-solving team.

However, regardless of the importance of good laws, fair-minded courts, brilliant executive leadership, and law-abiding behavior by law enforcement and the intelligence community, everyday people hold the key to changing the status quo to protect personal privacy in the new media environment.

It is simply unreasonable to assume that either government or commercial industries will make significant improvements in the privacy environment as long as citizens act as though they do not care about personal privacy. Opinion polls and preference surveys matter little when people vote with their hands, eyes, ears, and wallets. Each and every time Americans share personal information on social media; each time they report their location by using smartphones to log onto the Internet when they are away from home or their office; whenever otherwise free citizens use new media resources carelessly, they vote in favor of governments and companies continuing to collect, share, leverage, profit from, and otherwise use and abuse private personal data.

People who say "it's just advertising, I can ignore it" are correct. Generally, they can ignore the advertising. People who say "I'm just letting my friends and family know what is happening in my life—we are staying connected with Facebook" are indicating that they believe they are using the technology for the purpose that it was intended. When consumers answer surveys, return registration cards from retail purchases, "like" their favorite radio station, television station, magazine, or program on Facebook, or follow the cast and production team on Twitter, they are right: they are just being fans and it doesn't much matter. Those activities are what McLuhan would have labeled *content*. Content isn't the most important part of the message; the *medium is the message*.

What *really* matters is the infrastructure that's being built that can someday be used against Americans and democracy. What matters is that 1,000 times a day people act as though they do not care about personal privacy, even though if you ask them about it, very often they will say that aspects of privacy are still important to them. Not only do their usage habits cast a vote in favor of not caring, but those habits are also providing the building blocks for the construction of the infrastructure of surveillance. Remember that network functions may be the most important aspect in new media. The network not only connects, it builds exponentially. Add a node and you don't just add one (person or device); you add every person and/or device the person you added is connected to. Networks scale such that single actions taken by one person implicate countless connectivities. When you access the Web from your smartphone, the ads you see don't matter as much as the connection you are making with that device, as well as the data you contribute to the system at the same time.

Taking steps to more strongly protect one's personal privacy can be complex, difficult, and frustrating. The steps that one takes will not fully protect one's privacy; professional new media developers are always going to be a number of iterations ahead of lay citizens. Technology is developed and implemented by a very large number of very smart technology specialists. Currently, too many of them work against the best interest of their fellow citizens. If some act on the recommendations made in this chapter, recovering privacy will be less arduous than it is now.

Regardless of the difficulty, each time an individual makes a conscious effort to enact a particular behavior that goes against the grain and protects one's privacy, they are casting a vote for privacy protection for the rest of us and against the surveillance state. Changes in the current environment will only happen after citizens cast sufficient numbers of action-votes (decisions to protect their personal privacy) to get the attention of legislators and/or business interests. The market will only work to correct the privacy environment when consumers act as though they not only care, but care a lot.

Jolie O'Dell wrote in 2013 about "stopping the cycle" on *Venture Beat*, one of the hippest technology blogs in the business. Under the headline "Here's How to Stop the Gov't, NSA, FBI, and Everyone Else from Spying on You," she advised readers to "Log off. Opt out. And don't look back." More specifically, O'Dell proposed that

> If you actually care about government surveillance, the very first step you should take is stopping your use of social media. This is not hard to do. You can go totally cold turkey and barely notice it. If you really want to get serious about counter-surveillance, you can use Tor, Hushmail, and other such technologies. Again, these are designed to not be difficult. If you want to go all the way, you can ditch your smartphone. This takes some adjustment, but it's absolutely possible to live an organized life without a bleeping, blooping connection to all the worst parts of the Internet. (O'Dell 2013)

For most folks, it is not quite that simple. I will propose a number of strategies for user actions that both support personal privacy and send a message to commercial and governmental entities.

My treatment of ways that individuals can increase their personal privacy in the new media environment begins with a call for consciousness-raising: *Educate yourself about privacy and about how new media technologies work.* I begin with recommendations taken from four important sources beyond the current volume.

Michael Bazzell's *Hiding from the Internet: Eliminating Personal Online Information* (2013) strikes me as the single best resource for the wide-ranging activities that would be accomplished by someone wanting to come as close to 'getting off the grid' as is humanly possible in this day and age. Bazzell probably goes further than most folks are willing to venture. Nevertheless, the book is essential reading for the hyper-serious privacy advocate and provides numerous helpful specifications that can be used by folks who are seeking less dramatic changes.

The appendix, "Take Control of Your Data," in Adam Tanner's *What Stays in Vegas: The World of Personal Data—Lifeblood of Big Business—And the End of Privacy as We Know It* (2014) provides excellent palliative suggestions across a wide range of practices that generate and share personal data.

Likewise, Bernard E. Harcourt devotes two chapters of his book *Exposed: Desire and Disobedience in the Digital Age* (2015) to a section entitled "Digital Disobedience." These chapters are titled "Digital Resistance" and "Political Disobedience."

Two chapters in Julia Angwin's *Dragnet Nation: A Quest for Privacy, Security, and Freedom in a World of Relentless Surveillance* describe her efforts to assess the state of her personal privacy health ("Threat Models," "The Audit"). Note her efforts to assess her personal situation; emulate those efforts, and then recalibrate your behavior accordingly.

In addition to the current volume and the books just listed, see the Privacy Rights Clearinghouse (www.privacyrights.org) for privacy papers and information; the Electronic Privacy Information Center (epic.org/privacy/tools.html) for a list of privacy tools; and the Electronic Frontier Foundation's "I Fight Surveillance" website (https://ifightsurveillance.org/) for action-step recommendations and help guides.

Proposals for wide-ranging opposition to the status quo brought about by dominant media were not born yesterday. Gareth Branwyn's 1997 book, *Jamming the Media: A Citizen's Guide: Reclaiming the Tools of Communication*, was targeted primarily at loosening the strangleholds of old/mass media. However, while targeting "the mainstream media's monopoly on ideas," the book presents new media as the way to introduce new voices (1997: 16). Branwyn's book serves as an example of the early enthusiasm for new media technology's disruptive capabilities. The suggestions that he makes, although often targeted toward old/mass media, can now be used to remediate shortcomings in the new media infrastructure.

Any campaign against domination can benefit from a thorough reading of Saul Alinsky's *Rules for Radicals: A Practical Primer for Realistic Radicals* (1971).

Although Alinsky's work was targeted at community organizers, particularly those working in low-income and racially suppressed urban areas, his tactics, strategies, and rules are beneficial to oppositional forces wherever and whoever they may be.

Citizens can improve their privacy and encourage commercial and government entities to shift the status quo toward privacy protection across a number of genres/actions: credit and financial services, websites and browsers, social media participation, mobile phones, email, IoT, political action.

Individual actions (even when small) will improve the situation for citizens and the environment; wider-ranging, stronger actions could help change the status quo. Harcourt proposes strong actions as targets and goals, suggesting we

> scrutinize our personal information through self-help, education and awareness campaigns, and research leading to the development of more secure devices. The thrust here is to encrypt better, to create our own personal information management systems, to assemble our own secure servers, to retain our own information, to prevent exposing the data on the cloud. (2015: 270)

I offer the following recommendations. Taking these actions will help protect personal privacy and will send the message that users want businesses and government to do likewise.

- **Credit and Financial Services**
  - Free annual credit reports are available through a dedicated website: www.annualcreditreport.com. DO NOT go to the credit bureau websites; they make strong efforts to engage you in subscription services. Experian, TransUnion, and Equifax are the three businesses required to provide free annual reports; they are also on the list of the top 10 data mining companies (though you can't get data marketplace reports from them).
  - Are there times when you can use cash instead of credit? Separating your purchases from credit cards and mobile devices puts less information about your retail actions into the system.
  - Are there places where you use loyalty cards that are not of sufficient benefit to justify all the data they leak? Using loyalty cards guarantees total tracking and full entry of your actions into the data marketplace.
  - The major credit card companies enable you to opt out of some information sharing. See the privacy policies at their websites.

- Sign up for broad removal from data mining via links at StopData-Mining.me (http://www.stopdatamining.me/opt-out-list). Lifehacker links to a Reddit list of major public record/background check sites that you can remove your name from (http://lifehacker.com/5827106/remove-yourself-from-all-background-check-websites-a-master-list).

- **Websites and Browsers**
  - Whenever possible, access the World Wide Web from computers rather than from mobile devices, being especially careful to do as much of your Web work as possible from home or office (fixed locations). Mobile Web work exposes your data to networks that are not secure. Pushing Web work through a mobile device, especially your phone, often brings additional network providers into the mix, again lowering your control over settings and surveillance.
  - Use multiple browsers for specific purposes. Manage the security settings for those browsers depending on usage needs.
  - Reject third-party cookies as often as possible. Clear out cookies and browser caches often.
  - Remember that private browsing mode does not protect you or your machine from cookies and bugs.
  - Learn to use software that promotes anonymity, such as Tor.
  - There are reasonably priced devices and services that can mask your IP address location and other details (SurfEasy, Anonymizer.co) for particularly sensitive Web work.
  - Use Browser plug-ins that enable control over tracking and cookies (Blur, Disconnect.me, HTTPS Everywhere).
  - Sign up at the Digital Advertising Alliance page to opt out of targeted marketing/advertising functions (http://www.digitaladvertisingalliance.org/principle/index.html).

- **Social Media**
  - There's no nuanced way to use social media and protect your privacy. If you want to protect your privacy, and defend democracy, cancel your accounts and remove as much of your private information as you can.
  - Are you benefiting enough from your participation in social media to justify the time you put in? Are you treating everything you put into social media as information that you want to be shared with everyone

in the world? Every time you contribute to social media, businesses in the data marketplace make money—you don't. Are you tired of being shortchanged yet?
- Each of the social networks offers so-called "privacy controls." These enable you to block other members of the service, and Web-surfers in general, from seeing aspects of the information you provide. Setting these preferences can limit the views of other citizens. These settings *do not* change the ways that the social media provider and its many partners (and the data marketplace in general) treat your information.
- Privacychoice.org offers a usable dashboard in addition to controls available through the social media providers.
- Put information that you want to share with friends and family on a website that you control. Many ISPs and Web service providers offer low-cost sites featuring easy-to-use development tools *and* options for extensive privacy controls. Lock your site down from Web-crawlers, bugs, and search engines. Use the site to communicate with people you know; don't use the site to get famous on the Web or to make money. Limit access via log-ons that you set up for friends and family.
- If you insist on maintaining social media accounts, post minimal information there. Remember, however, that maintaining the registration connects others through you, via network effects. You may think that you are just a lurker or a benign observer; instead, you are a carrier, a conduit, a network connector as well. Starving the system is good; cutting yourself out of the system is better.

- **Mobile Phones**
  - Use your cell/mobile devices for phone calls.
  - Text messaging produces an enormous amount of unprotected data. Turning speech into text and sending it over mobile devices is as bad, or worse, for privacy than is posting to social media.
  - Federal law protects your privacy when using a wired phone (but not a wireless handset that is connected to a wired phone line). Reestablish landline services *now*; providers of those services are beginning to roll them back/cancel them, based on customer behavior (replacing landlines with cell phones).
  - Do not use your mobile/cell phone to access the Internet or any Web-based services. You pay for the convenience of watching video,

listening to music, surfing the Web (including using social media), and playing games with the diminution of your privacy.
    o Research and purchase secure communication devices such as the Blackphone, developed by geeksphone. Send the market the message that you want privacy-protected communication devices.

- **Electronic Mail**
    o Although it is possible to obtain 'privacy-responsible' email services, doing so is somewhat difficult and is exceedingly atypical.
    o Most email now circulates through the Google infrastructure.
    o Encrypted services are becoming progressively easier to use and, in terms of privacy protection (and keeping your email out of the data marketplace), they are well worth any trouble they put you through.
    o Setting up an account and using the systems is no longer the problem. However, getting your contacts on board with using encrypted mail can present serious roadblocks. If you alone have encrypted email, it doesn't help much; if you and the people you exchange mail with use encrypted email, it pretty much fixes the privacy problems with email.
    o Proton Mail may be the best current alternative. It is easy to install, is super easy to use, and raises the security of your email exponentially.
    o Transient email services, such as 10 Minute Mail, enable relatively anonymous email functionality that can disconnect your communication actions from your private data. For example, 10 Minute Mail enables consumers to sign up for products and services using a valid email address that 'disappears' after the transaction.

- **Internet of Things/Everything (IoT)**
    o Consider carefully the benefits and costs related to wearable monitoring devices. For instance, due to a malfunction in my cardio-electrical systems, I wear an implanted pacemaker. After nine years, the battery in my first device wore out. My cardiologist ordered a new device; it came with full wireless chip technology—the device's reporting functions (via an radio frequency chip—RF) were 'always on.' The device supplier promoted the RF features by noting that I would merely need to place a reception tower at my bedside so that the device could report my monitored condition regularly, especially at night. I specifically rejected that model and asked for a device without an RF chip; the device I wear now reports data only on my command, using specific

technology (a wired phone and the device supplier's transmitter) for that communication activity. The device is less convenient, but way less invasive, and my cardio-telemetry is not available for wireless/RF interception. I send reports, from my home, using a wired phone and a portable device reader that transmits when I enact the protocol.
- o Remember that the companies that provide personal monitors, and the networks they use, may not be secure and are not constrained by HIPAA policies or laws.
- o You will have to make decisions about how many 'smart devices' you are willing to allow into your life, home, car, and office. IoT devices might save money and may make your life more efficient. They will also generate enormous amounts of data for the data marketplace to turn into profit—profit that probably far offsets the savings that you see in your monthly bills. If one takes the electronic components currently inside a car as an example, we can speculate that IoT devices will not lower costs to consumers (although the auto industry claims they will).
- o In some cases (public utilities, for example), you may not be able to resist implementation of the IoT. Learn everything you can about how installed devices function and take active steps to insist that your information be constrained to the specific product/service. Often, IoT devices that go into the infrastructure are the most ignored, as consumers have to work hard (harder than they usually are willing to work) to ascertain the specifics of privacy policies and ToS.
- o Purchase non-IoT products whenever they are an option. Again, send producers/suppliers/sellers/industries the strongest message you can: Buy and use only those IoT products that meet your needs, provide demonstrated advantages, *and* feature clear and effective privacy protection. Avoid IoT devices that you don't really need or benefit from.

- **Political Action**
  - o If you are politically active, get involved with issues of privacy.
  - o Do not use social media for this function. Feeding the beast fattens it rather than producing starvation.
  - o Discipline the watchers by watching them. Work to render the internal operations of government, corporations, and the secret apparatus itself more visible. Again, do not use social media as the mechanism here.

- Use technology to perform actions that confuse the data marketplace and/or that render your data less (or not) valuable. For example, Adblock plus + AdNauseam sends more clicks to websites than you actually enter; TrackMeNot sends multiple search requests, even though you've just made one request. Technology can be used, legally, to confuse data analytics.
- Contact elected officials and demand that they legislate in support of privacy.
- Contact owners/operators of local organizations (commercial and/or nonprofit); examine their privacy policies and ToS. Provide constructive feedback indicating support for privacy protection.
- Write letters to the editor of your local newspaper. Provide constructive feedback indicating support for privacy protection.

In the final analysis, reestablishing personal privacy as a societal norm and priority requires action on the part of people in all societal segments. There aren't easy solutions. New media products and services are diligently and professionally designed, engineered, and implemented to be easy to use (frictionless) and highly rewarding. Nobody likes technology that is difficult to use. We do, however, need to take a more careful look than usual at the reward/punishment equation.

We are losing control over more and more parts of our everyday lives by the minute. For some, the deal is already done: Privacy is dead. Despite the ubiquitous evidence in favor of that conclusion, and the enormous difficulties associated with resisting that end state, I continue to be hopeful that Americans will wake up and smell the coffee and realize that information isn't, and doesn't want to be, free. The costs are much higher than most people know or are willing to admit.

Generations of Americans have given their lives in defense of democracy. Someone in your family probably went to war, and perhaps was wounded or killed. People you know and love sacrificed almost beyond measure so that we can live in a free, democratic society. You are being asked to log off Facebook, to put down your smartphone, to stop texting instead of talking, to read the fine print. In comparison to the sacrifices made by previous generations, the costs to you seem very pale.

The risks and punishments, however, are neither slim nor weak—nor speculative. The government collects and analyzes massive amounts of our personal data; commercial operations have amassed digital dossiers on each of us; the data marketplace bombards us with targeted marketing and advertising that becomes more personalized and more invasive every day. The infrastructure for surveillance and control is being built, click by click. And we are, generally, fully complicit. Democracy is not designed to operate within a surveillance state. Each of us can play a role in protecting our freedom. The Fair Information Practices remind us how.

NOTES

Chapter 3: FIP 1: No Secret Data Collection

1. Chapter 5 examines the third FIP, putting data to use as intended/provided.
2. Chapter 7 presents the outcome of my complaints over these forms and practices. UnityPoint Health–Methodist clarified its policies and agreed to modify its forms and paperwork.

Chapter 7: Recommendations

1. I took a group of undergraduate students to San Jose, California, for a week in the summer of 2014, hosted by our friends at Cisco Inc. The students worked on a project developed for them by Cisco and thereby learned a lot about new media workplaces. We are grateful to our friends Susan Monce and Christian Grossmann, Susan's supervisor Mike Mitchell, and the company, for their patience with us and for hosting us. I spent the week talking with folks at the company and discovered that "it is what it is" applies to many everyday situations in Cisco's corporate culture.

WORKS CITED

Abelson, Harold, Ross Anderson, Steven M. Bellovin, Josh Benaloh, Matt Blaze, Whitfield Diffie, John Gilmore, Matthew Green, Susan Landau, Peter G. Neumann, Ronald L. Rivest, Jeffrey I. Schiller, Bruce Schneier, Michael Specter, and Daniel J. Weitzner. "Keys under Doormats: Mandating Insecurity by Requiring Government Access to All Data and Communications." *DSpace@MIT*, July 6, 2015. <http://hdl.handle.net/1721.1/97690>. WEB 1 March, 2016.

Achenbach, Joel. "Meet the Digital Dissenters: They're Fighting for a Better Internet." *Washington Post*, December 27, 2015. <http://www.washingtonpost.com/sf/national/2015/12/26/resistance/>. WEB 27 December, 2015.

Ackerman, Spencer. "Fisa Chief Judge Defends Integrity of Court over Verizon Records Collection." *Guardian*, June 6, 2013. <http://www.theguardian.com/world/2013/jun/06/fisa-court-judge-verizon-records-surveillance>. WEB 6 November, 2015.

——— and Danny Yadron. "FBI Escalates War with Apple: 'Marketing' Bigger Concern than Terror." *Guardian*, February 19, 2016. <https://www.theguardian.com/technology/2016/feb/19/fbi-apple-san-bernardino-shooter-court-order-iphone>. WEB 28 June, 2016.

"Affinity Analysis." *Wikipedia*. Last modified May 21, 2015. <https://en.wikipedia.org/wiki/Affinity_analysis>. WEB 16 March, 2016.

Aftergood, Steven. "Privacy and the Imperative of Open Government." *Privacy in the Modern Age: The Search for Solutions*. Ed. Marc Rotenberg, Julia Horwitz, and Jeramie Scott. New York: The Free Press, 2015. 19–26. PRINT.

"A. G. Schneiderman Announces Settlement with Uber to Enhance Rider Privacy." Media release. New York State Office of the Attorney General. January 6, 2016. <http://www.

ag.ny.gov/press-release/ag-schneiderman-announces-settlement-uber-enhance-rider-privacy>. WEB 21 January, 2016.

Alinsky, Saul D. *Rules for Radicals: A Pragmatic Primer for Realistic Radicals*. New York: Random House, 1971. PRINT.

Allmer, Thomas, Christian Fuchs, Verena Kreilinger, and Sebastian Sevignani. "Social Networking Sites in the Surveillance Society: Critical Perspectives and Empirical Findings." *Media, Surveillance and Identity: Social Perspectives*. Ed. André Jansson and Miyase Christensen. New York: Peter Lang Publishers, 2012. 49–70. PRINT.

Amatriain, Xavier and Justin Basilico. "Netflix Recommendations: Beyond the 5 Stars (Part 1)." *Netflix Tech Blog*, April 6, 2012. <http://techblog.netflix.com/2012/04/netflix-recommendations-beyond-5-stars.html>. WEB 16 March, 2016.

____. Netflix Recommendations: Beyond the 5 Stars (Part 2)." *Netflix Tech Blog*, June 20, 2012. <http://techblog.netflix.com/2012/06/netflix-recommendations-beyond-5-stars.html>. WEB 16 March, 2016.

"Amicus Briefs in Support of Apple." *Apple Press Info*, n.d. <https://www.apple.com/pr/library/2016/03/03Amicus-Briefs-in-Support-of-Apple.html>. WEB 13 April, 2016.

Amway Global v. Woodward, 744 F. Supp. 657 (E.D. Mich. 2010).

Anderson, Janna and Lee Rainie. *Digital Life in 2025*. Pew Research Center. March 11, 2014. <http://www.pewinternet.org/2014/03/11/digital-life-in-2025/>. WEB 7 July, 2015.

Andrejevic, Mark. *Infoglut: How Much Information Is Changing the Way We Think and Know*. New York: Routledge, 2013. PRINT.

Angwin, Julia. *Dragnet Nation: A Quest for Privacy, Security, and Freedom in a World of Relentless Surveillance*. New York: Times Books, Henry Holt & Company, 2014. PRINT.

____. "The Web's New Gold Mine: Your Secrets." *Wall Street Journal*, July 30, 2010. <http://www.wsj.com/articles/SB10001424052748703940904575395073512989404>. WEB 12 July, 2015.

____. "Why Online Tracking Is Getting Creepier." *ars technica*, June 12, 2014. <http://arstechnica.com/tech-policy/2014/06/why-online-tracking-is-getting-creepier/>. WEB 12 July, 2015.

Apuzzo, Matt. "Should the Authorities Be Able to Access Your iPhone?" *New York Times*, February 17, 2016. <http://www.nytimes.com/2016/02/18/us/politics/whether-phones-should-lock-out-the-fbi.html>. WEB 22 February, 2016.

____. "WhatsApp Encryption Said to Stymie Wiretap Order." *New York Times*, March 12, 2016. <http://www.nytimes.com/2016/03/13/us/politics/whatsapp-encryption-said-to-stymie-wiretap-order.html>. WEB 13 April, 2016.

____, David E. Sanger, and Michael S. Schmidt. "Apple and Other Tech Companies Tangle with US over Data Access." *New York Times*, September 7, 2015. <http://www.nytimes.com/2015/09/08/us/politics/apple-and-other-tech-companies-tangle-with-us-over-access-to-data.html>. WEB 7 September, 2015.

Arias v. Intermex Wire Transfer, LLC. United States District Court, E.D. California Case No. 1:15-cv-01101 September 14, 2015.<http://www.leagle.com/decision/In%20FDCO%2020150915688/ARIAS%20v.%20INTERMEX%20WIRE%20TRANSFER,%20L.L.C.>. WEB 25 April, 2016.

Article 29 Data Protection Working Party (Art. 29 WP). "Appendix: List of Possible Compliance Measures." Ref. Ares (2014)3113072, September 23, 2014. <http://ec.europa.eu/justice/data-protection/article-29/documentation/other-document/files/2014/20140923_letter_on_google_privacy_policy_appendix.pdf>. WEB 28, January, 2016.

"Association Rule Learning." *Wikipedia*. Last modified June 10, 2016. <https://en.wikipedia.org/wiki/Association_rule_learning>. WEB 11 July, 2016.

"A Summary of Your Rights under the Fair Credit Reporting Act." Federal Trade Commission. n.d. <https://www.consumer.ftc.gov/sites/default/files/articles/pdf/pdf-0096-fair-credit-reporting-act.pdf>. WEB 27 January, 2016.

Auletta, Ken. *Googled: The End of the World as We Know It*. New York: Penguin Books, 2009. PRINT.

Bachman, Katy. "Poll: Targeted Advertising Is Not the Bogeyman; Nearly 70% Like at Least Some Tailored Internet Ads." *Adweek*, April 18, 2013. <http://www.adweek.com/news/technology/poll-targeted-advertising-not-bogeyman-updated-148649>. WEB 18 March, 2016.

Bamford, James. "NSA Is Building the Country's Biggest Spy Center." *Wired.com*, March 16, 2012. <http://www.wired.com/threatlevel/2012/03/ff_nsadatacenter/>. WEB 29 September, 2015.

____. "The Silent War," *Wired*, July, 2013. 90–99. PRINT.

____. "Who's in Big Brother's Database?" *New York Review of Books*, November 5, 2009. <http://www.nybooks.com/articles/archives/2009/nov/05/whos-in-big-brothers-database/>. WEB 7 October, 2015.

Barnes, Julian E. "US Military Plugs into Social Media for Intelligence Gathering." *Wall Street Journal*, August 6, 2014. <http://www.wsj.com/articles/u-s-military-plugs-into-social-media-for-intelligence-gathering-1407346557>. WEB 27 February, 2016.

Barrett, Brian. "Spotify Clears Up Its Controversial Privacy Policy." *Wired.com*, August 21, 2015. <http://www.wired.com/2015/08/spotify-clears-up-its-privacy-policy/>. WEB 31 August, 2015.

Bazzell, Michael. *Hiding from the Internet: Eliminating Personal Online Information*. Charleston, SC: CCI Publishing, 2013. PRINT.

Beckett, Lois. "Everything We Know about What Data Brokers Know about You." *ProPublica*, June 13, 2014. <https://www.propublica.org/article/everything-we-know-about-what-data-brokers-know-about-you>. WEB 7 February, 2016.

Benkler, Yochai. *The Wealth of Networks: How Social Production Transforms Markets and Freedom*. New Haven, CT: Yale University Press, 2006. PRINT.

Benner, Katie and Nicole Perlroth. "How Tim Cook, in iPhone Battle, Became a Bulwark for Digital Privacy." *New York Times*, February 18, 2016. <http://www.nytimes.com/2016/02/19/technology/how-tim-cook-became-a-bulwark-for-digital-privacy.html>. WEB 19 February, 2016.

Bielsa, Alberto. "Smart Agriculture Project in Galicia to Monitor Vineyards with Waspmote." *Libelium World*, June 8, 2012. <http://www.libelium.com/smart_agriculture_vineyard_sensors_waspmote>. WEB 16 March, 2016.

Big Data: Seizing Opportunities, Preserving Values. Executive Office of the President. May, 2014. <https://www.whitehouse.gov/sites/default/files/docs/big_data_privacy_report_may_1_2014.pdf>. WEB 1 May, 2014.

"Blizzard Entertainment Online Privacy Policy." Blizzard Entertainment. Last modified July 24, 2015. <http://us.blizzard.com/en-us/company/about/privacy.html> WEB 11 July, 2016.

boyd, danah, Karen Levy, and Alice Marwick. "The Networked Nature of Algorithmic Discrimination." *Data and Discrimination: Collected Essays*. Ed. Seeta Peña Gangadharan, Virginia Eubanks, and Solon Barocas. Washington, DC: New America Foundation Open Technology Institute, October, 2014. 53–57. <https://www.newamerica.org/downloads/OTI-Data-an-Discrimination-FINAL-small.pdf>. WEB 25 March, 2016.

Branwyn, Gareth. *Jamming the Media: A Citizen's Guide: Reclaiming the Tools of Communication*. San Francisco: Chronicle Books, 1997. PRINT.

Brill, Julie. "Demanding Transparency from Data Brokers." *Washington Post*, August 15, 2013. <https://www.washingtonpost.com/opinions/demanding-transparency-from-data-brokers/2013/08/15/00609680-0382-11e3-9259-e2aafe5a5f84_story.html>. WEB October 15, 2015.

Brin, Sergey and Lawrence Page. "The Anatomy of a Large-Scale Hypertextual Web Search Engine." Paper submitted to the Computer Science Dept., Stanford University, 1998. <http://infolab.stanford.edu/pub/papers/google.pdf>. WEB 17 March, 2016.

Broadband Consumer Privacy Proposal Fact Sheet. Federal Communications Commission. March 10, 2016. <https://www.fcc.gov/document/broadband-consumer-privacy-proposal-fact-sheet>. WEB 11 March, 2016.

Brodkin, Jon. "AT&T Accused of Violating Privacy Law with Sale of Phone Records to CIA." *ars technica*, December 11, 2013. <http://arstechnica.com/tech-policy/2013/12/att-accused-of-violating-privacy-law-with-sale-of-phone-records-to-cia/>. WEB 17 December, 2013.

———. "AT&T's Plan to Watch Your Web Browsing—And What You Can Do about It." *ars technica*, March 27, 2015. <http://arstechnica.com/information-technology/2015/03/27/atts-plan-to-watch-your-web-browsing-and-what-you-can-do-about-it/>. WEB 30 April, 2015.

———. "Report: Apple Designing Its Own Servers to Avoid Snooping." *ars technica*, March 24, 2016. <http://arstechnica.com/information-technology/2016/03/report-apple-designing-its-own-servers-to-avoid-snooping/>. WEB 4 April, 2016.

Brown, Eryn. "Digital Health Records Are Not Safe, Report on Breaches Shows." *Los Angeles Times*, April 14, 2014. <http://www.latimes.com/science/sciencenow/la-sci-sn-medical-records-breaches-20150414-story.html>. WEB 5 November, 2015.

Bureau of Justice Statistics. "16.6 Million People Experienced Identity Theft in 2012." Media release. Bureau of Justice Statistics. December 12, 2013. <http://www.bjs.gov/content/pub/press/vit12pr.cfm>. WEB 6 January, 2016.

Cairncross, Frances. *The Death of Distance: How the Communication Revolution Is Changing Our Lives*. Boston: Harvard Business School Press, 1997. PRINT.

"Call to Action." Decentralized Web Summit website, 2016. <http://www.decentralizedweb.net/>. WEB 8 June, 2016.

Calo, Ryan. "Digital Market Manipulation." *George Washington Law Review* 82.4 (2014): 996–1051. <http://www.gwlr.org/wp-content/uploads/2014/10/Calo_82_41.pdf>. WEB 10 October, 2014.

Carr, Nicholas. *The Glass Cage: Automation and Us*. New York: W. W. Norton & Company, 2014. PRINT.

Carrns, Ann. "The Cost to Consumers of a Data Breach." *New York Times*, April 30, 2013. <http://bucks.blogs.nytimes.com/2013/04/30/the-cost-to-consumers-of-a-data-breach/>. WEB 6 January, 2016.

Catacchio, Chad. "Some Reasons More People Aren't Using Check-In Services." *The Next Web News*, June 15, 2010. <http://thenextweb.com/location/2010/06/15/some-reasons-more-people-arent-using-check-in-services/>. WEB 20 July, 2010.

Centre for Economics and Business Research. *Data Equity: Unlocking the Value of Big Data*. Report for SAS. April, 2012. <https://www.sas.com/offices/europe/uk/downloads/data-equity-cebr.pdf>. WEB 20 January, 2016.

Cha, Ariana Eunjung. "The Revolution Will Be Digitized." *Washington Post*, May 9, 2015. <http://www.washingtonpost.com/sf/national/2015/05/09/the-revolution-will-be-digitized/>. WEB 13 May, 2016.

Channick, Robert. "Hey, You in the Altima! Chevy Billboard Spots Rival Cars, Makes Targeted Pitch." *Chicago Tribune*, April 15, 2016. <http://www.chicagotribune.com/business/ct-interactive-billboards-0417-biz-20160415-story.html>. WEB 15 April, 2016.

Christou v. Beatport, LLC, 849 F. Supp. 2d 1055 (D. Colo. 2012).

Cieply, Michael. "WikiLeaks Posts Sony Pictures Documents, Angering the Studio." *New York Times*, April 16, 2015. <http://www.nytimes.com/2015/04/17/business/media/sony-pictures-is-angered-by-wikileaks-posting-of-its-stolen-documents.html>. WEB 29 December, 2015.

Clifford, Stephanie. "Web Start-Ups Offer Bargains for Users' Data." *New York Times*, May 30, 2010. <http://www.nytimes.com/2010/05/31/business/media/31privacy.html>. WEB 31 May, 2010.

Cohn, Cindy and Kurt Opsahl. "Justice Delayed: Ninth Circuit Sends EFF's NSL Cases Back for Consideration under USA FREEDOM." Electronic Freedom Foundation. August 31, 2015. <https://www.eff.org/deeplinks/2015/08/justice-delayed-ninth-circuit-sends-effs-nsl-cases-back-consideration-under-usa>. WEB 10 January, 2016.

Cole v. Gene by Gene, Ltd., No. 1:14-cv-00004 (Alaska Dist. Ct., May 13, 2014).

"Commerce Control List (CCL)." Bureau of Industry and Security, US Dept. of Commerce, n.d. <http://www.bis.doc.gov/index.php/regulations/commerce-control-list-ccl/>. WEB 1 March, 2016.

Consumer Data Privacy in a Networked World: A Framework for Protecting Privacy and Promoting Innovation in the Global Digital Economy. Office of the President. February, 2012. <https://www.whitehouse.gov/sites/default/files/privacy-final.pdf>. WEB 2 July, 2015.

"Cookies and Other User and Ad-Targeting Technologies." *CNN.com*, last modified July 31, 2015. <http://www.cnn.com/privacy#turner_cookies>. WEB 17 March, 2016.

"Cows Can Text with M2M." *Telekom*, October 19, 2012. <http://www.telekom.com/media/enterprise-solutions/157120>. WEB 16 March, 2016.

Cox, Kate. "Supreme Court to Decide If You Can Sue When Data Aggregators Are Wrong." *Consumerist*, April 27, 2015. <http://consumerist.com/2015/04/27/supreme-court-to-decide-if-you-can-sue-when-data-aggregators-are-wrong/>. WEB 25 September 2015.

Crain, Matthew. *The Revolution Will Be Commercialized: Finance, Public Policy, and the Construction of Internet Advertising*. Doctoral dissertation, University of Illinois, Urbana-Champaign, 2013. PRINT.

Crane, Matthew. "US Export Controls on Technology Transfers." *Duke Law and Technology Review* 1.1 (2001). <http://scholarship.law.duke.edu/cgi/viewcontent.cgi?article=1029&context=dltr>. WEB 28 February, 2016.

"Criminal Attacks Are Now Leading Cause of Data Breach in Healthcare, According to New Ponemon Study." Media release. Ponemon Institute. May 7, 2015. <http://www.ponemon.org/news-2/66>. WEB 5 January, 2016.

Davis, Wendy. "Dispelling the Myth of 'Anonymous' Data." *MediaPost.com*, August 12, 2009. <http://www.mediapost.com/publications/article/111574/>. WEB 25 September, 2015.

———. "Do Users Read Privacy Policies? Answer Could Help Decide WikiLeaks/Twitter Case." *Daily Online Examiner*, March 28, 2011. <http://www.mediapost.com/publications/article/147591/do-users-read-privacy-policies-answer-could-help.html>. 27 November, 2014.

———. "FCC Proposes Broadband Privacy Rules." *Daily Online Examiner*, March 31, 2016. <http://www.mediapost.com/publications/article/272536/fcc-proposes-broadband-privacy-rules.html>. WEB 4 April, 2016.

Degeler, Andrii. "13 Million Mackeeper Users Exposed after Mongodb Door Was Left Open." *ars technica*, December 15, 2015. <http://arstechnica.com/security/2015/12/13-million-mackeeper-users-exposed-after-mongodb-door-was-left-open/>. WEB 21 December, 2015.

DiGangi, Christine. "These Identity Theft Statistics Are Even Scarier than You'd Expect." *Daily Finance*, December 31, 2013. <http://www.dailyfinance.com/2013/12/31/scariest-identity-theft-statistics/>. WEB 5 January, 2016.

Directive 95/46/Ec of the European Parliament and of the Council of 24 October, 1995 on the Protection of Individuals with Regard to the Processing of Personal Data and on the Free Movement of Such Data. <http://ec.europa.eu/justice/policies/privacy/docs/95-46-ce/dir1995-46_part1_en.pdf>. WEB 11 July, 2016.

"Director Discusses Encryption, Patriot Act Provisions." Federal Bureau of Investigation. May 20, 2015. <https://www.fbi.gov/news/news_blog/director-discusses-encryption-patriot-act-provisions>. WEB 13 April, 2016.

Doe v. Southeastern Pennsylvania Transportation Authority (SEPTA), 72 F.3d 1133 (3d Cir. 1995).

Dwoskin, Elizabeth. "Beware the Fine Print: FTC Settles Medical Data Case." *Wall Street Journal Law Blog*, December 3, 2014. <http://blogs.wsj.com/law/2014/12/03/beware-the-fine-print-ftc-settles-medical-data-case/>. WEB 5 November, 2015.

———. "FTC Recommends Limits on Data Collection via Internet of Things." *Wall Street Journal Law Blog*, January 27, 2015. <http://blogs.wsj.com/law/2015/01/27/ftc-recommends-limits-on-data-collection-via-internet-of-things/>. WEB 18 March, 2015.

"Edward Snowden Joins Reddit AMA." *Free Snowden*, February 23, 2015. <https://edwardsnowden.com/2015/02/24/edward-snowden-joins-reddit-ama/>. WEB 2 March, 2016.

Ehrenburg, Billy. "How Much Is Your Personal Data Worth?" *Guardian*, April 14, 2014. http://www.theguardian.com/news/datablog/2014/apr/22/how-much-is-personal-data-worth>. WEB 22 January, 2016.

Eisler, Peter and Susan Page. "3 NSA Veterans Speak Out on Whistle-Blower: We Told You So." *USA Today*, June 16, 2013. <http://www.usatoday.com/story/news/politics/2013/06/16/snowden-whistleblower-nsa-officials-roundtable/2428809/>. WEB 1 April, 2016.

Elgan, Mike. "Is Facial Recognition a Threat on Facebook and Google?" *Computerworld*, June 29, 2015. <http://www.computerworld.com/article/2941415/data-privacy/is-facial-recognition-a-threat-on-facebook-and-google.html>. WEB 29 June, 2015.

Essers, Loek. "Google Gets Privacy Lesson from EU Data Protection Authorities." *Computerworld*, September 26, 2014. <http://www.computerworld.com/article/2688001/google-gets-privacy-lesson-from-eu-data-protection-authorities.html>. WEB 4 November, 2014.

Estavillo v. Sony Computer Entertainment America, 2009 WL 3072887 (N.D. Cal., September 22, 2009).

Eubanks, Virginia. "Big Data and Human Rights." *Data and Discrimination: Collected Essays.* Ed. Seeta Peña Gangadharan, Virginia Eubanks, and Solon Barocas. Washington, DC: New America Foundation Open Technology Institute, October, 2014. 48–52. <https://www.newamerica.org/downloads/OTI-Data-an-Discrimination-FINAL-small.pdf>. WEB 25 March, 2016.

"Experts Predict Security and Privacy Trends for 2016." *id experts*, December 8, 2015. <https://www2.idexpertscorp.com/blog/single/top-four-privacy-and-security-predictions-for-2016>. WEB 28 March, 2016.

"Export Control." Office of Research Compliance, Ohio State University, n.d. <http://orc.osu.edu/regulations-policies/exportcontrol/>. WEB 28 February, 2016.

"Fact Sheet 7: Workplace Privacy and Employee Monitoring." Privacy Rights Clearinghouse. Last modified January, 2016. <https://www.privacyrights.org/workplace-privacy-and-employee-monitoring>. WEB 7 February, 2016.

Fainaru, Steve and James V. Grimaldi. "FBI Knew Terrorists Were Using Flight School." *Washington Post*, September 23, 2001. <http://www.prisonplanet.com/fbi_knew_terrorists_using_flight_schools.html>. WEB 7 April, 2016.

Fair and Accurate Credit Transactions Act of 2003 (FACTA), Pub. Law No. 108-159, 117 Stat. 1952 (December 4, 2003). <https://www.gpo.gov/fdsys/pkg/PLAW-108publ159/pdf/PLAW-108publ159.pdf>. WEB 25 January, 2016.

Fair Credit Reporting Act (FCRA). 15 U.S.C. § 1681. Federal Trade Commission. September, 2012. <https://www.consumer.ftc.gov/sites/default/files/articles/pdf/pdf-0111-fair-credit-reporting-act.pdf>. WEB 25 January, 2016.

Farivar, Cyrus. "County Sheriff Has Used Stingray Over 300 Times with No Warrant." *ars technica*, May 24, 2015. <http://arstechnica.com/tech-policy/2015/05/county-sheriff-has-used-stingray-over-300-times-with-no-warrant/>. WEB 14 June, 2015.

____. "Encrypted WhatsApp Messages Frustrate New Court-Ordered Wiretap." *ars technica*, March 12, 2016. <http://arstechnica.com/tech-policy/2016/03/encrypted-whatsapp-voice-calls-frustrate-new-court-ordered-wiretap/>. WEB 13 March, 2016.

____. "Even Former Heads of NSA, DHS Think Crypto Backdoors Are Stupid." *ars technica*, July 27, 2015. <http://arstechnica.com/tech-policy/2015/07/even-former-heads-of-nsa-dhs-think-crypto-backdoors-are-stupid/>. WEB 24 August, 2015.

____. "Even Former NSA Chief Thinks USA Freedom Act Was a Pointless Change." *ars technica*, June 17, 2015. <http://arstechnica.com/tech-policy/2015/06/even-former-nsa-chief-thinks-usa-freedom-act-was-a-pointless-change/>. WEB 23 June, 2015.

____. "FTC Asked to Block Uber from Getting Location Data in Background." *ars technica*, June 22, 2015. <http://arstechnica.com/tech-policy/2015/06/ftc-asked-to-block-uber-from-getting-location-data-in-background/>. WEB 23 June, 2015.

____. "Lawsuit Alleges Unauthorized Publication of Personal Genetics Data." *ars technica*, May 14, 2014. <http://arstechnica.com/tech-policy/2014/05/lawsuit-alleges-unauthorized-publication-of-personal-genetics-data/>. WEB 15 December, 2014.

____. "Online Behavioral Tracking Up 400 Percent across Prominent Websites." *ars technica*, June 18, 2012. <http://arstechnica.com/business/2012/06/online-behavioral-tracking-up-across-prominent-websites/>. WEB 19 June, 2012.

____. "Private Firms Sue Arkansas for Right to Collect License Plate Reader Data." *ars technica*, June 11, 2014. <http://arstechnica.com/tech-policy/2014/06/private-firms-sue-arkansas-for-right-to-collect-license-plate-reader-data/>. WEB 5 November, 2015.

____. "Secret Court Argues (Again) that It's Not a Rubber Stamp for Surveillance." *ars technica*. October 16, 2013. <http://arstechnica.com/tech-policy/2013/10/secret-court-argues-again-that-its-not-a-rubber-stamp-for-surveillance/>. WEB 6 November, 2015.

____. "Senator: Let's Fix 'Third-Party Doctrine' that Enabled NSA Mass Snooping." *ars technica*, April 4, 2016. <http://arstechnica.com/tech-policy/2016/04/senator-lets-fix-third-party-doctrine-that-enabled-nsa-mass-snooping/. WEB 6 April, 2016.

____. "Sheriff: We'll Get Judicial Approval—Not a Warrant—When Using Stingray." *ars technica*, September 18, 2015. <http://arstechnica.com/tech-policy/2015/09/sheriff-well-get-judicial-approval-not-a-warrant-when-using-stingray/>. WEB 23 September, 2015.

____. "Verizon Says It Received Over 321,000 Legal Orders for User Data in 2013." *ars technica*, January 22, 2014. <http://arstechnica.com/tech-policy/2014/01/verizon-says-it-received-over-321000-legal-orders-for-user-data-in-2013/>. WEB 6 November, 2015.

____. "We Know Where You've Been: Ars Acquires 4.6M License Plate Scans from the Cops." *ars technica*, March 24, 2015. <http://arstechnica.com/tech-policy/2015/03/we-know-where-youve-been-ars-acquires-4-6m-license-plate-scans-from-the-cops/>. WEB 25 March, 2015.

"FCC Adopts Strong, Sustainable Rules to Protect the Open Internet." Media release. Federal Communications Commission. February 26, 2015. <http://transition.fcc.gov/Daily_Releases/Daily_Business/2015/db0226/DOC-332260A1.pdf>. WEB 31 January, 2016.

"FDA Launches OpenFDA to Provide Easy Access to Valuable FDA Public Data." Media release. Food and Drug Administration. June 2, 2014. <http://www.fda.gov/NewsEvents/Newsroom/PressAnnouncements/ucm399335.htm>. WEB 5 November, 2015.

Fernandes, Teresa and Ana Calamite. "Unfairness in Consumer Services: Outcomes of Differential Treatment of New and Existing Clients." *Journal of Retailing and Consumer Services* 28 (January 2016): 36–44. <http://www.sciencedirect.com/science/article/pii/S096969891530028X>. WEB 22 March, 2016.

Finnegan, Leah. "William Cronon, University of Wisconsin Professor, Targeted by State Republicans." *Huffpost College*, March 28, 2011. <http://www.huffingtonpost.com/2011/03/28/william-cronon-university_n_841541.html>. WEB 6 March, 2016.

Fioretti, Julia, Shadia Nasralla, and Francois Murphy. "Max Schrems: The Law Student Who Took On Facebook." *Reuters*, October 7, 2015. <http://www.reuters.com/article/us-eu-ireland-privacy-schrems-idUSKCN0S124020151007>. WEB 5 February, 2016.

Fiveash, Kelly. "Google Extends Right-to-Be-Forgotten Rules to All Search Sites." *ars technica*, March 7, 2016. <http://arstechnica.com/tech-policy/2016/03/google-finally-extends-right-to-be-forgotten-rules-to-all-search-sites-including-com>. WEB 4 July, 2016.

"Former CIA Agent Says Edward Snowden Revelations Emboldened Apple to Push Back against FBI." *Democracy Now*, February 25, 2016. <http://www.democracynow.org/2016/2/25/former_cia_agent_says_edward_snowden>. WEB 3 March, 2016.

Frank, Blair Hanley. "Microsoft Doesn't See Windows 10's Mandatory Data Collection as a Privacy Risk." *Computerworld*, October 23, 2015. <http://www.computerworld.com/article/2996504/data-privacy/microsoft-doesnt-see-windows-10s-mandatory-data-collection-as-a-privacy-risk.html>. WEB 23 October, 2015.

Frankel, Todd C. "Apple's Tim Cook Calls FBI Request to Crack Terrorist's iPhone 'Bad for America.'" *Washington Post*, February 24, 2016. <https://www.washingtonpost.com/news/the-switch/wp/2016/02/24/apples-tim-cook-calls-fbi-request-to-crack-terrorists-iphone-bad-for-america/>. WEB 25 February, 2016.

Fried, Charles. "Privacy [A Moral Analysis]." *Yale Law Journal* 77 (1968): 21. PRINT.

Frye, Curt. "Up and Running with Public Data Sets." *Lynda.com*, June 1, 2015. <http://www.lynda.com/Tableau-tutorials/Up-Running-Public-Data-Sets/368761-2.html>. WEB 4 March, 2016.

Fung, Brian. "Internet Providers Want to Know More about You than Google Does, Privacy Groups Say." *Washington Post*, January 20, 2016. <https://www.washingtonpost.com/news/the-switch/wp/2016/01/20/your-internet-provider-is-turning-into-a-data-hungry-tech-company-consumer-groups-warn/>. WEB 21 January, 2016.

____. "Third-Party Cable Boxes Won't Be Allowed to Spy on You (Too Much), Regulators Vow." *Washington Post*, February 10, 2016. <http://wapo.st/1Q6ZlmU>. WEB 3 March, 2016.

____. "Why Civil Rights Groups Are Warning against 'Big Data'." *Washington Post*, February 27, 2014. <https://www.washingtonpost.com/news/the-switch/wp/2014/02/27/why-civil-rights-groups-are-warning-against-big-data/>. WEB 1 March, 2016.

Gallagher, Sean. "Director of National Intelligence: Snowden Forced 'Needed Transparency.'" *ars technica*. September 17, 2015. <http://arstechnica.com/tech-policy/2015/09/director-of-national-intelligence-snowden-forced-needed-transparency/>. WEB 30 September, 2015.

____. "FBI Says Crypto Ransomware Has Raked in >$18 Million for Cybercriminals." *ars technica*, June 25, 2015. <http://arstechnica.com/security/2015/06/fbi-says-crypto-ransomware-has-raked-in-18-million-for-cybercriminals/>. WEB 28 March, 2016.

____. "How the NSA Collects Millions of Phone Texts a Day." *ars technica*, November 9, 2014. <http://arstechnica.com/information-technology/2014/01/how-the-nsa-collects-millions-of-phone-texts-a-day/>. WEB 6 November, 2015.

____. "Machine Consciousness: Big Data Analytics and the Internet of Things. *ars technica*, March 24, 2015. <http://arstechnica.com/information-technology/2015/03/machine-consciousness-big-data-analytics-and-the-internet-of-things/>. WEB 15 June, 2015.

____. "When NSA and FBI Call for Surveillance Takeout, These Companies Deliver." *ars technica*, September 5, 2014. <http://arstechnica.com/tech-policy/2014/09/when-nsa-and-fbi-call-for-surveillance-takeout-these-companies-deliver/>. WEB 6 November, 2015.

Gangadharan, Seeta Peña. "Introduction." *Data and Discrimination: Collected Essays*. Ed. Seeta Peña Gangadharan, Virginia Eubanks, and Solon Barocas. Washington, DC: New America Foundation Open Technology Institute, October, 2014. 2–3. <https://www.newamerica.org/downloads/OTI-Data-an-Discrimination-FINAL-small.pdf>. WEB 25 March, 2016.

Gangadharan, Seeta Peña, Virginia Eubanks, and Solon Barocas (eds). *Data and Discrimination: Collected Essays*. Washington, DC: New America Foundation Open Technology Institute, October, 2014. <https://www.newamerica.org/downloads/OTI-Data-an-Discrimination-FINAL-small.pdf>. WEB 25 March, 2016.

Gatto, Mike. "Redcoat Solution to Government Surveillance." *Los Angeles Times*, September 29, 2015. <http://www.latimes.com/opinion/op-ed/la-oe-gatto-surveillance-3rd-amendment-20150929-story.html>. WEB 29 September, 2015.

Gellman, Barton and Laura Poitras. "US, British Intelligence Mining Data from Nine US Internet Companies in Broad Secret Program." *Washington Post*, June 7, 2013. <https://www.washingtonpost.com/investigations/us-intelligence-mining-data-from-nine-us-internet-companies-in-broad-secret-program/2013/06/06/3a0c0da8-cebf-11e2-8845-d970ccb04497_story.html>. WEB 27 February, 2016.

"Geo-Loco Conference: The Business & Future of Geo-Location Services." July 21, 2010. <https://www.eventbrite.com/e/geo-loco-conference-the-business-future-of-geo-location-services-tickets-544689180>. WEB 15 February, 2016.

Gershman, Jacob. "Court Orders Company to Make Website Accessible to the Blind." *Wall Street Journal Law Blog*, March 25, 2016. <http://blogs.wsj.com/law/2016/03/25/court-orders-company-to-make-website-accessible-to-the-blind>. WEB 4 April, 2016.

———. "Department of Justice and Twitter Clash over Disclosure of Surveillance Orders." *Wall Street Journal Law Blog*, March 3, 2016. <http://blogs.wsj.com/law/2016/03/03/department-of-justice-and-twitter-clash-over-disclosure-of-surveillance-orders/>. WEB 4 April, 2016.

———. "Judge: Browsing Data Shared with Facebook Doesn't Violate Privacy Law." *Wall Street Journal Law Blog*, April 10, 2015. <http://blogs.wsj.com/law/2015/04/10/judge-browsing-data-shared-with-facebook-doesnt-violate-privacy-law/>. WEB 12 April, 2015.

———. "Read the Ruling: Appeals Court Reverses on NSA Phone Surveillance." *Wall Street Journal Law Blog*, August 28, 2015. <http://blogs.wsj.com/law/2015/08/28/read-the-ruling-appeals-court-reverses-on-nsa-phone-surveillance/>. WEB 6 November, 2015.

Gidda, Mirren. "Edward Snowden and the NSA Files—Timeline." *Guardian*, August 21, 2013. <http://www.theguardian.com/world/2013/jun/23/edward-snowden-nsa-files-timeline>. WEB 7 October, 2015.

Goodin, Dan. "Apple Scrambles after 40 Malicious 'Xcodeghost' Apps Haunt App Store." *ars technica*, September 21, 2015. <http://arstechnica.com/security/2015/09/apple-scrambles-after-40-malicious-xcodeghost-apps-haunt-app-store/>. WEB 23 October, 2015.

———. "Chrome Extension Collects Browsing Data, Uses It for Marketing." *ars technica*, April 30, 2015. <http://arstechnica.com/security/2015/04/chrome-extension-collects-browsing-data-uses-it-for-marketing/>. WEB 31 August, 2015.

———. "Contrary to Public Claims, Apple Can Read Your iMessages." *ars technica*, October 17, 2013. <http://arstechnica.com/security/2013/10/contrary-to-public-claims-apple-can-read-your-imessages/>. WEB 22 October, 2013.

____. "Exposed: NSA Program for Hacking Any Cell Phone Network, No Matter Where It Is." *ars technica*, December 4, 2014. <http://arstechnica.com/tech-policy/2014/12/exposed-nsa-program-for-hacking-any-cellphone-network-no-matter-where-it-is/>. WEB 6 December, 2014.

____. "Researchers Find 256 iOs Apps that Collect Users' Personal Info." *ars technica*, October 20, 2015. <http://arstechnica.com/security/2015/10/researchers-find-256-ios-apps-that-collect-users-personal-info/>. WEB 23 October, 2015.

____. "Wish List App from Target Springs a Major Personal Data Leak." *ars technica*, December 15, 2015. <http://arstechnica.com/security/2015/12/wish-list-app-from-target-springs-a-major-personal-data-leak/>. WEB 21 December, 2015.

Gottsegen, Gordon. "You Can't Do Squat about Spotify's Eerie New Privacy Policy." *Wired.com*, August 20, 2015. <http://www.wired.com/2015/08/cant-squat-spotifys-eerie-new-privacy-policy/>. WEB 31 August, 2015.

Government Data Collection and Dissemination Practices Act (GDCDPA), Va. Code, § 2.2-3800 (2016). <http://law.lis.virginia.gov/vacode/2.2-3800/>. WEB 10 April, 2016.

Green, Shane. "What Is Your Personal Data Really Worth?" *TeamData Blog* [formerly Personal.com], February 13, 2012. <https://teamdata.com/blog/2012/02/what-is-your-personal-data-really-worth/>. WEB 22 January, 2016.

Greenslade, Roy. "How Edward Snowden Led Journalist and Film-Maker to Reveal NSA Secrets." *Guardian*, August 19, 2013. <http://www.theguardian.com/world/2013/aug/19/edward-snowden-nsa-secrets-glenn-greenwald-laura-poitras>. WEB 1 October, 2015.

Gross, Grant. "House Okays Cyberthreat Sharing Bill Despite Privacy Concerns." *Computerworld*, April 22, 2015. <http://www.computerworld.com/article/2913872/cyberwarfare/house-okays-cyberthreat-sharing-bill-despite-privacy-concerns.html>. WEB 6 November, 2015.

____. "Myspace Settles FTC Privacy Charges." *Computerworld*, May 8, 2012. <http://www.computerworld.com/article/2504027/technology-law-regulation/myspace-settles-ftc-privacy-charges.html>. WEB 8 May, 2012.

____. "US Privacy Watchdog: NSA Phone Records Program Is Illegal." *Computerworld*, January 23, 2014. <http://www.computerworld.com/article/2486904/government-it/u-s--privacy-watchdog--nsa-phone-records-program-is-illegal.html>. WEB 6 November, 2015.

Gunnarsson, Helen W. "Law and Technology: Your (Mostly Free) Private Investigator." *Illinois State Bar Association Journal* 90.4 (2010): 184. <https://www.isba.org/ibj/2010/04/yourmostlyfreeprivateinvestigator>. WEB 8 March, 2016.

Halbert, Jim and Thomas M. Boyd. "Analyzing the White House Consumer Privacy White Paper." *DLA Piper*, March 5, 2012. <https://www.dlapiper.com/en/us/insights/publications/2012/03/analyzing-the-white-house-consumer-privacy-white__/>. WEB 4 April, 2016.

Halzack, Sarah. "Target Data Breach Victims Could Get up to $10,000 Each from Court Settlement." *Washington Post*, March 19, 2015. <https://www.washingtonpost.com/news/business/wp/2015/03/19/target-data-breach-victims-could-get-up-10000-each-from-court-settlement/>. WEB 22 March, 2015.

Hamilton, Greg. "Why You Shouldn't Target Your Marketing: Target Marketing Fails." *MarketingSherpa Blog*, July 21, 2015. <http://sherpablog.marketingsherpa.com/marketing/target-marketing-fails/>. WEB 22 March, 2016.

Handler, Mark and Dalia Sussman. "Poll Finds Strong Acceptance for Public Surveillance." *New York Times*, April 30, 2013. <http://www.nytimes.com/2013/05/01/us/poll-finds-strong-acceptance-for-public-surveillance.html>. WEB 30 April, 2013.

Harcourt, Bernard E. *Exposed: Desire and Disobedience in the Digital Age*. Cambridge, MA: Harvard University Press, 2015. PRINT.

Hardesty, Larry. "How Hard Is It to 'De-Anonymize' Cellphone Data?" *MIT News Office*, March 27, 2013. <https://news.mit.edu/2013/how-hard-it-de-anonymize-cellphone-data>. WEB 1 October, 2015.

Hardy, Quentin. "The Web's Creator Looks to Reinvent It." *New York Times*, June 7, 2016. <http://www.nytimes.com/2016/06/08/technology/the-webs-creator-looks-to-reinvent-it.html>. WEB 8 June, 2016.

Henton, Timothy J. "Interactive Social Media: The Value for Law Enforcement." *The FBI Enforcement Bulletin*, September 3, 2013. <https://leb.fbi.gov/2013/september/interactive-social-media-the-value-for-law-enforcement>. WEB 10 October, 2013.

"How Can Watson Health Help You?" IBM, n.d. <http://www.ibm.com/smarterplanet/us/en/ibmwatson/health/>. WEB 1 April, 2016.

"How Consumers Foot the Bill for Data Breaches." *Authentify.com*, August 8, 2014. <http://authentify.com/wp-content/uploads/2015/01/Authentify-Data-Breaches-Infographic.pdf>. WEB 17 March, 2016.

"How Search Works." Google, n.d. <https://www.google.com/insidesearch/howsearchworks/index.html>. WEB 17 March, 2016.

Hsu, Spencer C. "A Maryland Court Is the First to Require a Warrant for Covert Cellphone Tracking." *Washington Post*, March 31, 2016. <https://www.washingtonpost.com/world/national-security/a-maryland-court-is-the-first-to-require-a-warrant-for-covert-cellphone-tracking/2016/03/31/472d9b0a-f74d-11e5-8b23-538270a1ca31_story.html>. WEB 9 April, 2016.

Hsukayama, Hayley. "People Care More about Convenience than Privacy Online." *Washington Post*, October 7, 2014. <http://www.washingtonpost.com/blogs/the-switch/wp/2014/10/07/people-care-more-about-convenience-than-privacy-online/>. WEB 4 November, 2014.

"Identifying Encryption Items: What Items Are Removed from Encryption Controls?" Department of Industry and Security, US Department of Commerce, n.d. <http://www.bis.doc.gov/index.php/policy-guidance/encryption/identifying-encryption-items#Three>. WEB 3 March, 2016.

Internet of Things: Privacy & Security in a Connected World. Federal Trade Commission Staff Report, January, 2015. <https://www.ftc.gov/system/files/documents/reports/federal-trade-commission-staff-report-november-2013-workshop-entitled-internet-things-privacy/150127iotrpt.pdf>. WEB 17 March, 2016.

Isaac, Mike. "Apple Still Holds the Keys to Its Cloud Service, but Reluctantly." *New York Times*, February 21, 2016. <http://www.nytimes.com/2016/02/22/technology/apple-still-holds-the-keys-to-its-cloud-service-but-reluctantly.html>. WEB 22 February, 2016.

———. "Surge in Twitter Data Requests." *New York Times*, February 9, 2015. <http://bits.blogs.nytimes.com/2015/02/09/twitter-reports-surge-in-government-data-requests/>. WEB 10 February, 2015.

ITRC Breach Statistics 2005–2015. Identity Theft Resource Center. n.d. <http://www.idtheftcenter.org/images/breach/2005to2015multiyear.pdf>. WEB 28 March, 2016.

ITRC Data Breach Report. Identity Theft Resource Center. December 29, 2015. <http://www.idtheftcenter.org/images/breach/DataBreachReports_2015.pdf>. WEB 30 December, 2015.

ITRC Data Breach Report. Identity Theft Resource Center. March 22, 2016. <http://www.idtheftcenter.org/images/breach/DataBreachReports_2016.pdf>. WEB 27 March, 2016.

"Jacobs Calls on DMV to Stop Selling Drivers' Personal Information." Erie County Clerk's Office, New York. May 20, 2015. <http://www2.erie.gov/clerk/index.php?q=-jacobs-calls-dmv-stop-selling-drivers%E2%80%99-personal-information>. WEB 15 March, 2016.

Jansson, André and Miyase Christensen (eds). *Media, Surveillance and Identity: Social Perspectives.* New York: Peter Lang Publishers, 2014. PRINT.

Jenkins, Holman W. Jr. "Google and the Search for the Future." *Wall Street Journal*, August 14, 2010. <http://www.wsj.com/articles/SB10001424052748704901104575423294099527212>. WEB 22 March, 2016.

"*John Doe, a SEPTA Employee v. SEPTA and Pierce.*" 1995 Decisions. Paper 320. 1995. <http://digitalcommons.law.villanova.edu/cgi/viewcontent.cgi?article=1319&context=thirdcircuit_1995>. WEB 16 October 2015.

Jouvenal, Justin. "The New Way Police Are Surveilling You: Calculating Your Threat 'Score.'" *Washington Post*, January 10, 2016. <https://www.washingtonpost.com/local/public-safety/the-new-way-police-are-surveilling-you-calculating-your-threat-score/2016/01/10/e42bccac-8e15-11e5-baf4-bdf37355da0c_story.html>. WEB 11 January, 2016.

"Judgment in Case C-131/12 [*Google Spain SL, Google Inc. v Agencia Española de Protección de Datos, Mario Costeja González.*]." Press Release No. 70/14. Court of Justice of the European Union. May 13, 2014. <http://curia.europa.eu/jcms/upload/docs/application/pdf/2014-05/cp140070en.pdf>. WEB 5 February, 2016.

Kahneman, Daniel. *Thinking, Fast and Slow.* New York: Farrar, Straus & Giroux, 2011. PRINT.

Kang, Cecilia. "FCC Proposes Privacy Rules for Internet Providers." *New York Times*, March 10, 2016. < http://www.nytimes.com/2016/03/11/technology/fcc-proposes-privacy-rules-for-internet-providers.html>. WEB 11 March, 2016.

_____. "Verizon Settles with FCC over Hidden Tracking via 'Supercookies.'" *New York Times*, March 7, 2016. <http://www.nytimes.com/2016/03/08/technology/verizon-settles-with-fcc-over-hidden-tracking.html>. WEB 10 March, 2016.

Kehl, Danielle, Andi Wilson, and Kevin Bankston. "Doomed to Repeat History? Lessons from the Crypto Wars of the 1990s." Open Technology Institute, New America. June, 2015. <https://static.newamerica.org/attachments/3407--125/Lessons%20From%20the%20Crypto%20Wars%20of%20the%201990s.882d6156dc194187a5fa51b14d55234f.pdf>. WEB 1, March, 2016.

Keister, Lisa A. "The One Percent." *Annual Review of Sociology* 40 (2014):16.1–16.21. <https://wealthinequality.org/content/uploads/2015/10/Keister-The-One-Percent.pdf>. WEB 15 December, 2015.

Keizer, Gregg. "Apple Faces Questions from Congress about iPhone Tracking." *Computerworld*, April 21, 2011. <http://www.computerworld.com/article/2507868/data-privacy/apple-faces-questions-from-congress-about-iphone-tracking.html>. WEB 21 April, 2011.

Khandelwal, Swati. "Global Internet Authority—ICANN Hacked Again!" *thehackernews.com*, August 6, 2015. <https://thehackernews.com/2015/08/icann-hacked.html>. WEB 29 December, 2015.

Kihn, Martin. "Agenda Overview for Data-Driven Marketing, 2015." *Gartner*, December 29, 2014. <http://www.gartner.com/imagesrv/digital-marketing/pdfs/Data_Driven_Marketing.pdf>. WEB 22 March, 2016.

Kim, Nancy S. *Wrap Contracts: Foundations and Ramifications*. Oxford: Oxford University Press, 2013. PRINT.

Kirkpatrick, David. *The Facebook Effect: The Inside Story of the Company that Is Connecting the World*. New York: Simon & Schuster, 2010. PRINT.

Koh, Yoree. "Twitter Sues US Government over Data Requests." *Wall Street Journal Law Blog*, October 7, 2014. <http://blogs.wsj.com/law/2014/10/07/twitter-sues-u-s-government-over-data-requests/>. WEB 9 November, 2014.

Kravets, David. "FBI Embracing Facial Recognition to 'Find Bad Guys.'" *ars technica*, June 11, 2014. <http://arstechnica.com/tech-policy/2014/06/fbi-embracing-facial-recognition-to-find-bad-guys/>. WEB 15 December, 2014.

_____. "Google Has Removed 170,000-Plus URLs under 'Right to Be Forgotten' Edict." *ars technica*, October 10, 2014. <http://arstechnica.com/tech-policy/2014/10/google-has-removed-170000-plus-urls-under-right-to-be-forgotten-edict/>. WEB 9 November, 2014.

_____. "Google Rejecting 59 Percent of Right-to-Be-Forgotten Removal Requests." *ars technica*, May 15, 2013. <http://arstechnica.com/tech-policy/2015/05/google-rejecting-59-percent-of-right-to-be-forgotten-removal-requests/>. WEB 16 June, 2015.

_____. "Let the Snooping Resume: Senate Revives Patriot Act Surveillance Measure." *ars technica*, June 14, 2015. <http://arstechnica.com/tech-policy/2015/06/let-the-snooping-resume-senate-revives-patriot-surveillance-measures/>. WEB 6 November, 2015.

_____. "Surveillance Society Grows as Government Demands for Twitter Data Jump." *ars technica*, February 10, 2015. <http://arstechnica.com/tech-policy/2015/02/surveillance-society-grows-as-government-demands-for-twitter-data-jump/>. WEB 5 November, 2015.

_____. "Twitter Says Gag on Surveillance Scope Is Illegal 'Prior Restraint.'" *ars technica*, October 7, 2014. <http://arstechnica.com/tech-policy/2014/10/twitter-says-gag-on-surveillance-scope-is-illegal-prior-restraint/>. WEB 21 November, 2014.

_____. "UN Says Encryption 'Necessary for the Exercise of the Right to Freedom.'" *ars technica*, May 28, 2015. <http://arstechnica.com/tech-policy/2015/05/un-says-encryption-necessary-for-the-exercise-of-the-right-to-freedom/>. WEB 14 June, 2015.

_____. "Worker Fired for Disabling GPS App that Tracked Her 24 Hours a Day." *ars technica*, June 16, 2015. <http://arstechnica.com/tech-policy/2015/05/worker-fired-for-disabling-gps-app-that-tracked-her-24-hours-a-day/>. WEB 7 January, 2016.

Lamoureux, Edward Lee. "A Privacy Challenge for the Enterprise." *InterBusiness Issues*, April, 2014. <http://www.peoriamagazines.com/ibi/2014/apr/privacy-challenge-enterprise>. WEB 1 May, 2014.

____. "On Leaving Facebook: Confessions of a New Media Theorist and Teacher." *InterBusiness Issues*, October, 2010. <http://www.peoriamagazines.com/ibi/2010/oct/leaving-facebook>. WEB 1 November, 2010.

Landers, Peter. "You Be the Judge—Who's Right on NSA Data Gathering?" *Wall Street Journal*, December 30, 2013. <http://blogs.wsj.com/law/2013/12/27/you-be-the-judge-whos-right-on-nsa-data-gathering/>. WEB 5 October, 2015.

Lazarus, David. "Europe and US Have Different Approaches to Protecting Privacy of Personal Data." *Los Angeles Times*, December 22, 2015. <http://www.latimes.com/business/la-fi-lazarus-20151222-column.html>. WEB 27 January, 2016.

Lessig, Lawrence. *Code, Version 2.0*. New York: Basic Books, 2006. PRINT.

____. *The Future of Ideas: The Fate of the Commons in a Connected World*. New York: Random House, 2001. PRINT.

Levine, Robert, "Behind the European Privacy Ruling that's Confounding Silicon Valley." *New York Times*, October 9, 2015. <http://www.nytimes.com/2015/10/11/business/international/behind-the-european-privacy-ruling-thats-confounding-silicon-valley.html>. WEB 16 October, 2015.

Levy, Steven. *Crypto: How the Code Rebels Beat the Government—Saving Privacy In the Digital Age*. New York: Penguin Books, 2001. PRINT.

Lewis, Paul. "US Government Faces Pressure after Biggest Leak in Banking History." *Guardian*, February 8, 2015. <http://www.theguardian.com/news/2015/feb/08/us-government-biggest-leak-banking-history-questions-irs-taxes>. WEB 29 December, 2015.

Libert, Tim. "Health Privacy Online: Patients at Risk." *Data and Discrimination: Collected Essays*. Ed. Seeta Peña Gangadharan, Virginia Eubanks, and Solon Barocas. Washington, DC: New America Foundation Open Technology Institute, October, 2014. 11–15. <https://www.newamerica.org/downloads/OTI-Data-an-Discrimination-FINAL-small.pdf>. WEB 25 March, 2016.

"LifeLock to Pay $100 Million to Consumers to Settle FTC Charges It Violated 2010 Order." Media release. Federal Trade Commission. December 17, 2015. <https://www.ftc.gov/news-events/press-releases/2015/12/lifelock-pay-100-million-consumers-settle-ftc-charges-it-violated>. WEB 6 December, 2016.

"LifeLock Will Pay $12 Million to Settle Charges by the FTC and 35 States that Identity Theft Prevention and Data Security Claims Were False." Media release. Federal Trade Commission. March 9, 2010. <https://www.ftc.gov/news-events/press-releases/2010/03/lifelock-will-pay-12-million-settle-charges-ftc-35-states>. WEB 6 December, 2016.

Lohr, Steve. "Microsoft Sues US over Orders Barring It from Revealing Surveillance." *New York Times*, April 14, 2016. <http://www.nytimes.com/2016/04/15/technology/microsoft-sues-us-over-orders-barring-it-from-revealing-surveillance.html>. WEB 14 April, 2016.

Lynch, Jennifer. "FBI to Have 52 Million Photos in Its NGI Face Recognition Database by Next Year." *ars technica*, August 14, 2014. <http://arstechnica.com/tech-policy/2014/04/fbi-to-have-52-million-photos-in-its-ngi-face-recognition-database-by-next-year/>. WEB 7 December, 2014.

Madden, Mary. "Public Perceptions of Privacy and Security in the Post-Snowden Era." Pew Research Center. November 12, 2014. <http://www.pewinternet.org/2014/11/12/public-privacy-perceptions/>. WEB 25 November, 2014.

____, Amanda Lenhart, Sandra Cortesi, and Urs Gasser. "Teens and Mobile Apps Privacy." Pew Research Center. August 22, 2013. <http://www.pewinternet.org/2013/08/22/teens-and-mobile-apps-privacy/>. WEB 15 October, 2015.

Manick, Mike. "Candidate Obama Debating President Obama on Civil Liberties vs. Government Surveillance." *Tech Dirt*, June 21, 2013. <https://www.techdirt.com/articles/20130621/11024123555/candidate-obama-debating-president-obama-civil-liberties-vs-government-surveillance.shtml. WEB 5 April, 2016.

Maras, Elliot. "2015: The Year of the Breach; Close to 200 Million Personal Records Exposed." *Hacked*, December 15, 2015. <https://hacked.com/2015-year-breach-close-200-million-personal-records-exposed/>. WEB 30 December, 2015.

Markoff, John. "Cryptography Pioneers Win Turing Award." *New York Times*, March 1, 2016. <http://www.nytimes.com/2016/03/02/technology/cryptography-pioneers-to-win-turing-award.html>. WEB 1 March, 2016.

____, Katie Benner, and Brian X. Chen. "Apple Encryption Engineers, If Ordered to Unlock iPhone, Might Resist." *New York Times*, March 17, 2016. <http://www.nytimes.com/2016/03/18/technology/apple-encryption-engineers-if-ordered-to-unlock-iphone-might-resist.html>. WEB 12 April, 2016.

Marshall, Jack. "Do Consumers Really Want Targeted Ads?" *Wall Street Journal*, April 17, 2014. <http://blogs.wsj.com/cmo/2014/04/17/do-consumers-really-want-targeted-ads/>. WEB 18 March, 2016.

Masters, Greg. "On Deck: Predictions for 2016 and Beyond." *SC Magazine*, December 14, 2016. <http://www.scmagazine.com/on-deck-predictions-for-2016-and-beyond/article/458529/3/>. WEB 28 March, 2016.

McChesney, Robert W. *Digital Disconnect: How Capitalism Is Turning the Internet against Democracy*. New York: The Free Press, 2013. PRINT.

____. *Rich Media, Poor Democracy: Communication Politics in Dubious Times*. New York: The New Press, 2015. PRINT.

McCullagh, Declan. "Privacy Dispute Tests Obama's Earlier Promises." *CNET*, April 8, 2011. <http://www.cnet.com/news/privacy-dispute-tests-obamas-earlier-promises/>. WEB 5 April, 2016.

McFarland, Matt. "Why Facebook's Push to End Poverty Is Actually Self-Serving." *Washington Post*, February 16, 2016. <https://www.washingtonpost.com/news/innovations/wp/2016/02/16/when-facebook-gives-internet-to-the-poor-the-real-winner-isnt-whom-you-think/>. WEB 17 February, 2016.

McLuhan, Marshall and Quentin Fiore. *The Medium Is the Massage: An Inventory of Effects*. New York: Penguin Books, 1967. PRINT.

McMeley, Christin, John D. Seiver, Ronald London, and Bryan Thompson. "Supreme Court Grants Cert in *Spokeo v. Robins*." *Privacy and Security Law Blog*, April 27, 2015. <http://www.privsecblog.com/2015/04/articles/marketing-and-consumer-privacy/supreme-court-grants-cert-in-spokeo-v-robins/>. WEB 7 January, 2016.

McPhail, Michael J. "Categories of Hacker." Last updated October 14, 1996. <http://courses.cs.vt.edu/~cs3604/lib/Hacking/Landreth.html>. WEB 30 December, 2015.
———. "Donn Parker's Categories of Hacker." Last updated April 30, 1996. <http://courses.cs.vt.edu/~cs3604/lib/Hacking/Parker.html>. WEB 30 December, 2015.
———. "Hackers and Hacking, CS 3604 Participant, Fall 1996." Last updated July 1, 1998. <http://courses.cs.vt.edu/~cs3604/lib/Hacking/notes.html>. WEB 30 December, 2015.
McStay, Andrew. *Privacy and Philosophy: New Media and Affective Protocol*. New York: Peter Lang Publishers, 2014. PRINT.
Mearian, Lucas. "'Wall of Shame' Exposes 21M Medical Record Breaches." *Computerworld*, August 7, 2012. <http://www.computerworld.com/s/article/9230028/_Wall_of_Shame_exposes_21M_medical_record_breaches>. WEB 7 August, 2012.
Miller, Vincent. *The Crisis of Presence in Contemporary Culture: Ethics, Privacy, and Speech in the Mediated Social Life*. Los Angeles: Sage, 2016. PRINT.
Milligan, Robert B. "California Legislature Passes Bill to Extend Social Media Privacy Laws to Public Employers." *Trading Secrets Law Blog*, September 17, 2013. <http://www.tradesecretslaw.com/2013/09/articles/trade-secrets/california-legislature-passes-bill-to-extend-social-media-privacy-laws-to-public-employers/>. WEB 3 November, 2014.
———. "Michigan Adopts Social Media Privacy Legislation." *Trading Secrets Law Blog*, January 8, 2013. <http://www.tradesecretslaw.com/2013/01/articles/legislation-2/employers-take-note-michigan-adopts-social-media-privacy-legislation/>. WEB 3 November, 2014.
Mills, Jon L. *Privacy in the New Media Age*. Gainesville: University of Florida Press, 2015. PRINT.
Miners, Zach. "Google, Facebook, Microsoft Show Steady Rise in Surveillance Data." *Computerworld*, February 3, 2014. <http://www.computerworld.com/s/article/9245986/Google_Facebook_Microsoft_show_steady_rise_in_surveillance_data_requests>. WEB 4 February, 2014.
———. "Google Ordered to Remove Links to Stories about 'Right to Be Forgotten' Request." *Computerworld*, August 20, 2015. <http://www.computerworld.com/article/2974044/data-privacy/google-ordered-to-remove-links-to-right-to-be-forgotten-removal-stories.html>. WEB 21 August, 2015.
Moody, Glyn. "British Users Can Sue Google in UK over 'Secret Tracking.'" *ars technica*, March 27, 2015. <http://arstechnica.com/tech-policy/2015/03/27/british-users-can-sue-google-in-uk-over-secret-tracking/>. WEB 26 April, 2015.
Moore, Angela K. "Re: Compliance Concern." December 7, 2015. Personal letter to author.
Mullin, Joe. "How NSA Leaks Are Changing Minds among the Public—And in Congress." *ars technica*, July 30, 2013. <http://arstechnica.com/tech-policy/2013/07/how-nsa-leaks-are-changing-minds-among-the-public-and-in-congress/>. WEB 12 September, 2015.
Naeimi, Helia, Srinivasan Natarajan, Kushagra Vaid, Prabhakar Kudva, and Maitreya Natu. "Cloud Atlas—Unreliability through Massive Connectivity." IEEE VTS 31st VLSI Test Symposium, April 30, 2013, Berkeley, CA. DOI:10.1109/VTS.2013.6548907. WEB 22 March, 2016.
Nakashima, Ellen. "Judge Rules in Favor of Apple in Key Case Involving a Locked iPhone." *Washington Post*, February 29, 2016. <https://www.washingtonpost.com/world/national-

security/judge-rules-in-favor-of-apple-in-key-case-involving-a-locked-iphone/2016/02/29/fa76783e-db3d-11e5-925f-1d10062cc82d_story.html>. WEB 29 February, 2016.

———. "Justice Department: Agencies Need Warrants to Use Cellphone Trackers." *Washington Post*, September 4, 2015. <http://wapo.st/1QbytEx>. WEB 6 November, 2015.

———. "Top EU Court Strikes Down Major Data Sharing Pact between US and Europe." *Washington Post*, October 6, 2015. <http://wapo.st/1Q4DrTg>. WEB 10 October, 2015.

Narayan, Kaushik. "Dyre Straits: Why This Cloud Attack's Different." *Information Week*, September 12, 2014. <http://www.informationweek.com/cloud/software-as-a-service/dyre-straits-why-this-cloud-attacks-different/a/d-id/1315677>. WEB 27 March, 2016.

Narayanan, Arvind and Vitaly Shmatikov. "How to Break Anonymity of the Netflix Prize Dataset, v2." arXiv:cs/0610105v2 [cs.CR], November 22, 2007. <http://arxiv.org/abs/cs/0610105#>. WEB 25 September, 2015.

National Security Agency (NSA). "Mission and Strategy." Last modified May 3, 2016. <https://www.nsa.gov/about/mission-strategy>. WEB 12 July, 2016.

"National Security Letters Are Unconstitutional, Federal Judge Rules." Media release. Electronic Frontier Foundation. March 15, 2013. <https://www.eff.org/press/releases/national-security-letters-are-unconstitutional-federal-judge-rules>. WEB 16 March, 2013.

Negroponte, Nicholas P. *Being Digital*. New York: Alfred A. Knopf, 1995. PRINT.

Neuburger, Jeffrey. "Clickwrap Agreement Available Only through Hyperlink Enforceable under New York Law." *Proskauer: New Media and Technology Law Blog*, July 23, 2015. <http://newmedialaw.proskauer.com/2015/07/23/clickwrap-agreement-available-only-through-hyperlink-enforceable-under-new-york-law/>. WEB 2 October, 2015.

———. "Facial Recognition Technology: Social Media and Beyond, an Emerging Concern." *Proskauer: New Media and Technology Law Blog*, June 23, 2015. <http://newmedialaw.proskauer.com/2015/06/23/facial-recognition-technology-social-media-and-beyond-an-emerging-concern>. WEB 24 June, 2015.

New York Times Editorial Board. "The President on Mass Surveillance." *New York Times*, January 14, 2014. <http://nyti.ms/1axNazd>. WEB 18 January, 2016.

Nissenbaum, Helen. *Privacy in Context: Technology, Policy, and the Integrity of Social Life*. Stanford, CA: Stanford Law Books, 2010. PRINT.

———. "'Respect for Context': Fulfilling the Promise of the White House Report." *Privacy in the Modern Age*. Ed. Marc Rotenberg, Julia Horwitz, and Jeramine Scott. New York: The New Press, 2015. 152–164. PRINT.

Nixon, Ron. "US to Further Scour Social Media Use of Visa and Asylum Seekers." *New York Times*, February 23, 2016. <http://www.nytimes.com/2016/02/24/us/politics/homeland-security-social-media-refugees.html>. WEB 5 March, 2016.

Noyes, Katherine. "IBM's New $1B Acquisition Will Help Watson 'See,' Analyze Health Data." *Computerworld*, August 6, 2015. <http://www.computerworld.com/article/2960520/healthcare-it/ibms-new-1b-acquisition-will-help-watson-see-analyze-health-data.html>. WEB 6 August, 2015.

O'Brien, Kevin. "Silicon Valley Companies Lobbying against Europe's Privacy Proposals." *New York Times*, January 25, 2013. <http://nyti.ms/SNn03J>. WEB 26 January, 2013.

O'Dell, Jolie. "Here's How to Stop the Gov't, NSA, FBI, and Everyone Else from Spying on You." *Venture Beat*, November 18, 2013. <http://venturebeat.com/2013/11/18/heres-how-to-stop-the-govt-nsa-fbi-and-everyone-else-from-spying-on-you/>. WEB 5 March, 2016.

Olmstead v. United States, 277 U.S. 438 (1928).

Opsahl, Kurt. *Crypto Wars Part II: The Empires Strike Back.* Lecture, Chaos Communication Congress, Hamburg, Germany, December 30, 2015. Video recording. <https://www.youtube.com/watch?v=Z3a06NEq2Mc>. WEB 22 February, 2016.

"Opt Out List." *StopDataMining.me*, n.d. <http://www.stopdatamining.me/opt-out-list/>. WEB 1 April, 2016.

O'Reilly, Tim. "What Is Web 2.0.: Design Patterns and Business Models for the Next Generation of Software." *O'Reilly*, September 30, 2005. <http://www.oreilly.com/pub/a/web2/archive/what-is-web-20.html>. WEB 22 January, 2016.

Oremus, Will. "I Didn't Type This Article: Speech Recognition Has Gotten Better than You Think." *Slate*, April 23, 2014. <http://www.slate.com/articles/technology/technology2014/04the_end_of_typing_speech_recognition_technology_is_getting_better_and_better.html>. WEB 22 March, 2016.

Ornstein, Charles. "Abortion Foes Find New Ways to Get Details about Patients, Doctors." *Washington Post*, August 25, 2015. <http://wapo.st/1LvxNIz>. WEB 26 August, 2015.

———. "Federal Privacy Law Lags Far Behind Personal-Health Technologies." *Washington Post*, November 17, 2015. <http://wapo.st/1X4H6Ti>. WEB 17 November, 2015.

"Overview of the Privacy Act of 1974." Last modified July 16, 2015. <http://www.justice.gov/opcl/policy-objectives>. WEB 10 February, 2016.

"PageRank." *Wikipedia*, last modified July 8, 2016. <https://en.wikipedia.org/wiki/PageRank>. WEB 9 July, 2016.

Pagliery, Jose. "Wikileaks Publishes Searchable Database of Hacked Sony Emails." *CNN.com*, April 16, 2015. <http://money.cnn.com/2015/04/16/technology/security/wikileaks-sony-hack/index.html>. WEB 29 December, 2015.

Palazzolo, Joe. "Court Orders for Surveillance Hit Record in 2013." *Wall Street Journal Law Blog*, September 14, 2014. <http://blogs.wsj.com/law/2014/09/17/court-orders-for-surveillance-hit-record-high-in-2013/>. WEB 21 November, 2014.

Patient Authorization for Release of Information. UnityPoint Health–Methodist, Peoria, IL. Received by author, October 26, 2015.

Peacock, Sylvia E. "Prized Assets for Web Tracking." *SMSociety '15*. Proceedings of the 2015 International Conference on Social Media and Society, New York. DOI: 10.1145/2789187.2789199. WEB 22 March, 2016.

Peel, Deborah. "The Future of Health Privacy." *Privacy in the Modern Age*. Ed. Marc Rotenberg, Julia Horwitz, and Jeramine Scott. New York: The New Press, 2015. 174–180. PRINT.

Peterson, Andrea. "Consumer Group Wants Government to Make Google Give Americans the 'Right to Be Forgotten' Online." *Washington Post*, July 7, 2015. <http://wapo.st/1fj6F4J>. WEB 7 July, 2015.

———. "French Regulators Tell Google to Hide 'Right to Be Forgotten' Removals on All Sites." *Washington Post*, September 21, 2015. <http://wapo.st/1V6ouXg>. WEB 21 September, 2015.

——. "How the Country's Top Privacy Cop Is Trying to Protect Consumers in the Digital Age." *Washington Post*, June 6, 2015. <https://www.washingtonpost.com/news/the-switch/wp/2015/06/06/how-the-countrys-top-privacy-cop-is-trying-to-protect-consumers-in-the-digital-age/>. WEB 8 June, 2015.

——. "Massive Database of Over 190 Million Registered Voters' Information Leaked Online." *Washington Post*, December 29, 2015. <https://www.washingtonpost.com/news/the-switch/wp/2015/12/29/massive-database-of-over-190-million-registered-voters-information-leaked-online/>. WEB 29 December, 2015.

——. "Why It's So Hard to Keep Up with How the US Government Is Spying On Its Own People." *Washington Post*, November 20, 2015. <http://wapo.st/1kLGQgm>. WEB 20 November, 2015.

PhoneDog v. Kravitz, U.S. Dist. LEXIS 129229, 2011 WL 5415612 (N.D. Cal. November 8, 2011).

"Poll: Americans Want Free Internet Content, Value Interest-Based Advertising." *Digital Advertising Alliance*, April 18, 2013. <http://www.aboutads.info/DAA-Zogby-Poll>. WEB 18 March, 2016.

Ponemon Institute. "2015 Cost of Data Breach Study: United States." Ponemon Institute. May, 2015. <https://public.dhe.ibm.com/common/ssi/ecm/se/en/sew03055usen/SEW03055USEN.PDF>. WEB 6 January, 2016.

Ponemon, Larry. "Cost of Data Breaches Rising Globally, Says '2015 Cost of a Data Breach Study: Global Analysis.'" *Security Intelligence*, May 27, 2015. <https://securityintelligence.com/cost-of-a-data-breach-2015/>. WEB 5 January, 2016.

Postman, Neil. *Building a Bridge to the Eighteenth Century: How the Past Can Improve Our Future*. New York: Alfred A. Knopf, 1999. PRINT.

——. *Technopoly: The Surrender of Culture to Technology*. New York: Knopf, 1992. PRINT.

PricewaterhouseCoopers LLP. *IAB Internet Advertising Revenue Report: 2014 Full Year Results*. Interactive Advertising Bureau (IAB). April, 2015. <http://www.iab.com/wp-content/uploads/2015/05/IAB_Internet_Advertising_Revenue_FY_2014.pdf>. WEB 22 January, 2016.

Privacy and Behavioral Research [known as *The Hornig Report*]. Washington, DC: Office of Science and Technology of the Executive Office of the President, 1967. PRINT.

"Privacy Legislation: ALPRs, Stingrays, Drones, NSA." Tenth Amendment Center. n.d. <http://tenthamendmentcenter.com/legislation/privacy/>. WEB 12 April, 2016.

Privacy on the Go: Recommendations for the Mobile Ecosystem, California Department of Justice, Privacy Enforcement and Protection Unit. January, 2013. <http://oag.ca.gov/sites/all/files/agweb/pdfs/privacy/privacy_on_the_go.pdf>. WEB 22 March, 2016.

"Privacy Policy." Google. n.d. <https://www.google.com/intl/en/policies/privacy/>. WEB 31 August, 2015.

"Privacy Policy." Instagram. n.d. <https://instagram.com/about/legal/privacy/>. WEB 25 September, 2015.

"Privacy Policy." *NYTimes.com*, last updated June 10, 2015. <http://www.nytimes.com/content/help/rights/privacy/policy/privacy-policy.html>. WEB 25 September, 2015.

"Protecting and Promoting the Open Internet." FCC Rule. 80 Fed. Reg. 19737 (April 13, 2015). <https://federalregister.gov/a/2015-07841>. WEB 31 January, 2016.

Quibell, Marc. "NSA Surveillance Pales Compared to Social Media Data Collection." *Tripwire*, December 26, 2013. <http://www.tripwire.com/state-of-security/security-awareness/over-sharing-marc-quibell/>. WEB 3 March, 2016.

Quinn, Gene and Steve Brachmann. "The 'Right to Be Forgotten,' an EU Regulation Washing Up on American Shores." *IPWatchdog*, August 24, 2015. <http://www.ipwatchdog.com/2015/08/24/the-right-to-be-forgotten/id=60568/>. WEB 25 August, 2015.

Rainie, Lee. "13 Things to Know about Teens and Technology." Pew Research Center. July 23, 2014. <http://www.pewinternet.org/2014/07/23/13-things-to-know-about-teens-and-technology/>. WEB 16 October, 2015.

____ and Maeve Duggan. "Privacy and Information Sharing." Pew Research Center. January 14, 2016. <http://www.pewinternet.org/2016/01/14/privacy-and-information-sharing/>. WEB 8 February, 2016.

____ and Mary Madden. "Americans' Privacy Strategies Post-Snowden." Pew Research Center. March 16, 2015. <http://www.pewinternet.org/2015/03/16/americans-privacy-strategies-post-snowden/>. WEB 16 October, 2015.

Rall, Ted. "FBI vs. Apple Is Really about Snowden." *Counterpunch*, March 1, 2016. <http://www.counterpunch.org/2016/03/01/fbi-vs-apple-is-really-about-snowden/>. WEB 1 March, 2016.

Records, Computers, and the Rights of Citizens: Report of the Secretary's Advisory Committee on Automated Personal Data Systems. US Department of Health, Education, and Welfare. July, 1973. <https://epic.org/privacy/hew1973report/>. WEB 25 June, 2015.

Reeves, Byron and Clifford Nass. *The Media Equation: How People Treat Computers, Television, and New Media Like Real People and Places*. Palo Alto, CA: CSLI Publications, 1996. PRINT.

"Reform Government Surveillance: Read Our Open Letter to the Senate on the USA Freedom Act." *Reform Government Surveillance*, May 19, 2015. <https://www.reformgovernmentsurveillance.com>. WEB 20 February, 2016.

Richards, Amanda. "FTC Can Sue Companies with Poor Information Security, Appeals Court Says." *ars technica*, August 25, 2015. <http://arstechnica.com/tech-policy/2015/08/ftc-can-sue-companies-with-poor-information-security-appeals-court-says/>. WEB 25 August, 2015.

Richards, Neil M. "Dangers of Surveillance." *Harvard Law Review* 126 (2013):1–32. <http://ssrn.com/abstract=2239412>. WEB 23 April, 2013.

Rigley, Michael. *Network*. BFA thesis, Graphic Design, California College of Arts, 2011. <https://vimeo.com/34750078>. WEB 22 January, 2016.

Riley v. California, 134 S.Ct. 2473 (2014).

Risen, James and Laura Poitras. "NSA Collecting Millions of Faces from Web Images." *New York Times*, May 31, 2013. <http://www.nytimes.com/2014/06/01/us/nsa-collecting-millions-of-faces-from-web-images.html>. WEB 5 June, 2013.

____. "NSA Gathers Data on Social Connections of US Citizens." *New York Times*, September 28, 2013. <http://www.nytimes.com/2013/09/29/us/nsa-examines-social-networks-of-us-citizens.html>. WEB 5 October, 2013.

Robinson, Stephen Cory. "The Good, the Bad, and the Ugly: Applying Rawlsian Ethics in Data Mining Marketing." *Journal of Media Ethics* 30.1 (2015): 19–30. DOI:10.1080/08900523.2014.985297. WEB 13 March, 2106.

Rosen, Armin. "A New Report Paints an Alarming Picture about the Kind of Cybercrime You Rarely Hear About." *Business Insider*, February 6, 2016. <http://www.businessinsider.com/china-russia-hackers-cyber-criminals-2016-2>. WEB 7 February, 2016.

Rotenberg, Marc and Alan Butler. "In *Riley v. California*, a Unanimous Supreme Court Sets Out Fourth Amendment for Digital Age." *SCOTUSblog*, June 26, 2014. <http://www.scotusblog.com/2014/06/symposium-in-riley-v-california-a-unanimous-supreme-court-sets-out-fourth-amendment-for-digital-age/>. WEB 5 January, 2016.

____, Julia Horwitz, and Jeramie Scott (eds). *Privacy in the Modern Age: The Search for Solutions*. New York: The Free Press, 2015. PRINT.

Rubin, Howard and Don Stait. "Oregon Passes Social Media in the Workplace Law." *Littler*, June 6, 2013. <http://www.littler.com/oregon-passes-social-media-workplace-law>. WEB 3 November, 2014.

Ryan, Johnny. *A History of the Internet and the Digital Future*. London: Reaktion Press, 2010. PRINT.

Saavedra, Tony. "Watchdog: Can the Police Track Your Cellphone without You Knowing?" *Orange County Register*, March 11, 2015. <http://www.ocregister.com/articles/stingray-653962-aclu-police.html>. WEB 29 March, 2015.

Safire, William. "You Are a Suspect." *New York Times*, November 14, 2002. <http://www.nytimes.com/2002/11/14/opinion/14SAFI.html>. WEB 6 November, 2015.

Sapiezynski, Piotr, Arkadiusz Stopczynski, Radu Gatej, and Sune Lehmann. "Tracking Human Mobility Using WiFi Signals." *PLoS ONE* 10.7 (2015): e0130824. DOI:10.1371/journal.pone.0130824. WEB 7 January, 2016.

Sasso, Brendan. "FTC Chief Calls for 'Targeted' Regulation of Uber, Airbnb." *The Atlantic*, October 2, 2015. <http://www.theatlantic.com/politics/archive/2015/10/ftc-chief-calls-for-targeted-regulation-of-uber-airbnb/457458/>. WEB 4 April, 2016.

Savage, Charlie. "CIA Is Said to Pay AT&T for Call Data." *New York Times*, November 7, 2013. <http://www.nytimes.com/2013/11/07/us/cia-is-said-to-pay-att-for-call-data.html>. WEB 7 November, 2013.

____. "File Says NSA Found Way to Replace Email Program." *New York Times*, November 19, 2015. <http://www.nytimes.com/2015/11/20/us/politics/records-show-email-analysis-continued-after-nsa-program-ended.html>. WEB 21 November, 2015.

____. "NSA Said to Search Content of Messages to and from US." *New York Times*, August 8, 2013. <http://www.nytimes.com/2013/08/08/us/broader-sifting-of-data-abroad-is-seen-by-nsa.html>. WEB 7 October, 2015.

____ and Jonathan Weisman. "NSA Collection of Bulk Call Data Is Ruled Illegal." *New York Times*, May 7, 2015. <http://www.nytimes.com/2015/05/08/us/nsa-phone-records-collection-ruled-illegal-by-appeals-court.html>. WEB 6 November, 2015.

Savage, David G. "DC Appeals Court Lifts Injunction against NSA's Bulk-Collection Program." *Los Angeles Times*, August 28, 2015. <http://www.latimes.com/nation/la-na-court-nsa-bulk-collection-20150828-story.html>. WEB 28 April, 2016.

SB 236 Government Data Collection and Dissemination Practices Act (GDCDPA); Collection and Use of Personal Information. Introduced by J. Chapman Petersen, Va. Senate, January 13, 2016. <http://lis.virginia.gov/cgi-bin/legp604.exe?161+sum+SB236>. WEB 11 April, 2016.

Schaefers, Scott A. "Nevada and Colorado Pass Employee Social Networking Privacy Laws." *Trading Secrets Law Blog*, July 1, 2013. <http://www.tradesecretslaw.com/2013/07/articles/social-media-2/nevada-and-colorado-pass-employee-social-networking-privacy-laws/>. WEB 3 November, 2014.

———. "Washington State Passes Social Networking Privacy Legislation." *Trading Secrets Law Blog*, May 27, 2013. <http://www.tradesecretslaw.com/2013/05/articles/practice-procedure/washington-state-passes-social-media-privacy-legislation/>. WEB 3 November, 2014.

Scheer, Robert. *They Know Everything about You: How Data-Collecting Corporations and Snooping Government Agencies Are Destroying Democracy*. New York: Nation Books, 2015. PRINT.

Schneier, Bruce. "Fear and Convenience." *Privacy in the Modern Age: The Search for Solutions*. Ed. Marc Rotenberg, Julia Horwitz, and Jeramie Scott. New York: The Free Press, 2015. 200–203. PRINT.

Scola, Nancy. "Why Campaigns Won't Stop Using Our Data: Because the Data Says We Like It." *Washington Post*, December 18, 2014. <http://wapo.st/1r3Fr6x>. WEB 18 December, 2014.

"Secret." *Merriam-Webster.com*, n.d. <http://www.merriam-webster.com/dictionary/secret>. WEB 1 August, 2015.

Self-Regulatory Principles for Online Behavioral Advertising. American Association of Advertising Agencies, Association of National Advertisers, Council of Better Business Bureaus, Direct Marketing Association, and Interactive Advertising Bureau. July, 2009. <http://www.iab.net/media/file/ven-principles-07-01-09.pdf>. WEB 2 July, 2015.

Self-Regulatory Principles for Online Behavioral Advertising: Behavioral Advertising: Tracking, Targeting, and Technology. Federal Trade Commission Staff Report. February, 2009. <https://www.ftc.gov/sites/default/files/documents/reports/federal-trade-commission-staff-report-self-regulatory-principles-online-behavioral-advertising/p085400behavadreport.pdf>. WEB 2 July, 2015.

Shah, Mahek. "3 Technology Trends Transforming Health Care." *Forbes*, April 14, 2015. <http://www.forbes.com/sites/athenahealth/2015/04/13/3-technology-trends-transforming-health-care/#2816a0962601>. WEB 5 May, 2015.

"SilverHook Powerboats Develops a Racing App 40 Percent Faster." *IBM*, March, 2015. <http://www-01.ibm.com/software/ebusiness/jstart/portfolio/SilverHookCaseStudy.pdf>. WEB 16 March, 2016.

Silver, Joe. "Microsoft Challenges US Gov't Warrant to Access Overseas Customer Data." *ars technica*, June 11, 2014. <http://arstechnica.com/tech-policy/2014/06/microsoft-challenges-us-govt-warrant-to-access-overseas-customer-data/>. WEB 15 December, 2014.

Singel, Ryan. "Federal Efforts in Data Privacy Move Slowly." *New York Times*, February 28, 2016. <http://www.nytimes.com/2016/02/29/technology/obamas-effort-on-consumer-privacy-falls-short-critics-say.html>. WEB 29 February, 2016.

———. "Whistle-Blower Outs NSA Spy Room." *Wired*, April 7, 2006. <http://archive.wired.com/science/discoveries/news/2006/04/70619>. WEB 7 April, 2016.

Singer, Natasha. "If My Data Is an Open Book, Why Can't I Read It?" *New York Times*, May 25, 2013. <http://nyti.ms/12UHryv>. WEB 26 May, 2013.

Skloot, Rebecca. "Your Cells. Their Research. Your Permission?" *New York Times*, December 30, 2015. <http://www.nytimes.com/2015/12/30/opinion/your-cells-their-research-your-permission.html>. WEB 30 December, 2015.

"Social Media Is Part of Today's Workplace but Its Use May Raise Employment Discrimination Concerns." Media release. US Equal Employment Opportunity Commission. March 12, 2014. <http://www.eeoc.gov/eeoc/newsroom/release/3-12-14.cfm>. WEB 7 February, 2016.

"Social Media Use in Law Enforcement." *LexisNexis*, November, 2014. <http://www.lexisnexis.com/risk/elqNow/elqRedir.htm?ref=http://www.lexisnexis.com/risk/downloads/whitepaper/2014-social-media-use-in-law-enforcement.pdf>. WEB 1 March, 2016.

Solove, Daniel J. *Nothing to Hide: The False Tradeoff between Privacy and Security*. New Haven, CT: Yale University Press, 2011. PRINT.

_____. *The Digital Person: Technology and Privacy in the Information Age*. New York: New York University Press, 2004. PRINT.

"Spokeo, Inc. v. Robins." *Oyez*. Chicago-Kent College of Law at Illinois Tech, n.d. <https://www.oyez.org/cases/2015/13-1339>. WEB 22 January, 2016.

Spokeo, Inc. v. Robins. "Petition for a Writ of Certiorari." May 1, 2014. <http://sblog.s3.amazonaws.com/wp-content/uploads/2014/05/13-1339-Spokeo-v-Robins-Cert-Petition-for-filing.pdf>. WEB 22 January, 2016.

Staines, Rupert. "Consumers Have Accepted Advertising Online, but Targeted Mobile Ads? Not So Much." *NewStatesman*, February 4, 2014. <http://www.newstatesman.com/business/2014/02/consumers-have-accepted-advertising-online-targeted-mobile-ads-not-so-much>. WEB 18 March, 2016.

Staples, William G. *Everyday Surveillance: Vigilance and Visibility in Postmodern Life*, 2nd ed. Lanham, MD: Rowman & Littlefield, 2014. PRINT.

Storm, Darlene. "LA's Plan to Scan License Plates and Send Dear Prostitute-Seeking John Letters." *Computerworld*, December 2, 2015. <http://www.computerworld.com/article/3011394/security/las-plan-to-scan-license-plates-and-send-dear-prostitute-seeking-john-letters.html>. WEB 2 December, 2015.

Tanner, Adam. *What Stays in Vegas: The World of Personal Data—Lifeblood of Big Business—And the End of Privacy as We Know It*. New York: PublicAffairs/Perseus Books Group, 2014. PRINT.

Tene, Omer. "Privacy Law's Midlife Crisis: A Critical Assessment of the Second Wave of Global Privacy Laws." *Ohio State Law Journal* 74.6 (2013): 1218–1261. <https://kb.osu.edu/dspace/bitstream/handle/1811/71612/OSLJ_V74N6_1217.pdf>. WEB 4 April, 2016.

"Terms of Service." Google. n.d. <https://www.google.com/intl/en/policies/terms/>. WEB 28 July, 2015.

Thaler, Richard H. and Cass R. Sunstein. *Nudge: Improving Decisions about Health, Wealth, and Happiness*. New Haven, CT: Yale University Press, 2009.

"The 2010 Where 2.0 Conference Is a Wrap." *Where 2.0 Conference* website, n.d. <http://whereconf.com/where2010>. WEB 15 February, 2016.

Third Annual 2016 Data Breach Industry Forecast. Experian Data Breach Resolution, 2016. <http://www.experian.com/assets/data-breach/white-papers/2016-experian-data-breach-industry-forecast.pdf>. WEB 28 March, 2016.

Thompson, David Sr. "2016 Security Trends: What's Next for Data Breaches?" *IT BusinessEdge*, n.d. <http://www.itbusinessedge.com/slideshows/2016-security-trends-whats-next-for-data-breaches.html>. WEB 28 March, 2016.

Tsukayama, Hayley. "This Is How Much Money You're Worth to Facebook." *Washington Post*, January 27, 2016. <https://www.washingtonpost.com/news/the-switch/wp/2016/01/27/this-is-how-much-money-youre-worth-to-facebook/>. WEB 28 January, 2016.

———. "This Smart Car Seems to Have Tattled on Its Driver." *Washington Post*, December 7, 2015. <http://wapo.st/1NCh3ln>. WEB 7 December, 2015.

Turow, Joseph. *The Daily You: How the New Advertising Industry Is Defining Your Identity and Your Worth*. New Haven, CT: Yale University Press, 2011. PRINT.

"Uber Error Leaks US-Based Drivers' Data." *BBC*, October 14, 2015. <http://www.bbc.com/news/technology-34529821>. WEB 29 December, 2015.

United States v. Wurie (see *Riley v. California*).

US Public Record Databases. Gallagher Law Library, University of Washington School of Law. January 10, 2014. <https://lib.law.washington.edu/content/research/USPubRec>. WEB 2 March, 2016.

Vanderbilt, Tom. "The Science behind the Netflix Algorithms that Decide What You'll Watch Next." *Wired.com*, August 7, 2013. <http://www.wired.com/2013/08/qq_netflix-algorithm/>. WEB 17 March, 2016.

van Dijck, José. *The Culture of Connectivity: A Critical History of Social Media*. Oxford: Oxford University Press, 2014. http://www.uva.nl/over-de-uva/organisatie/medewerkers/content/d/i/j.f.t.m.vandijck/j.f.t.m.vandijck.html

Vidal-Hall & Ors v. Google Inc. [2014] E.W.H.C. 13 (Q.B.), January 16, 2014. <http://www.bailii.org/ew/cases/EWHC/QB/2014/13.html>. WEB 4 February, 2016.

Vijayan, Jaikumar. "BP Employee Loses Laptop Containing Data on 13,000 Oil Spill Claimants." *Computerworld*, March 29, 2011. <http://www.computerworld.com/article/2507424/data-security/bp-employee-loses-laptop-containing-data-on-13-000-oil-spill-claimants.html>. WEB 6 January, 2016.

Ward, Mark. "Cryptolocker Victims to Get Files Back for Free." *BBC News*. August 6, 2014. <http://www.bbc.com/news/technology-28661463>. WEB 28 March, 2016.

Warner, Bernhard. "What the Intelligence Community Is Doing with Big Data." *Bloomberg*, February 5, 2013. <http://www.bloomberg.com/bw/articles/2013-02-05/what-the-intelligence-community-is-doing-with-big-data>. WEB 10 February, 2013.

Warren, Samuel D. and Louis D. Brandeis. "The Right to Privacy." *Harvard Law Review* 4 (1890–91): 193–220. <http://louisville.edu/law/library/special-collections/the-louis-d.-brandeis-collection/the-right-to-privacy>. WEB 11 July, 2016.

Wertheimer, Michael. "The Mathematics Community and the NSA: Encryption and the NSA Role in International Standards." *Notices of the American Mathematical Society* 62.2 (2015): 165–167. DOI:10.1090/noti1213. WEB 15 January, 2015.

Westin, Alan F. *Privacy and Freedom*. New York: Atheneum, 1967. PRINT.

Whalen, Andrew. "New Barbie Doll Spying on Your Kids? Wi-Fi Connected Talking Hello Barbie Doll Linked to CIA Money." *iDigital Times*, November 4, 2015. <http://www.idigitaltimes.com/new-barbie-doll-spying-your-kids-wi-fi-connected-talking-hello-barbie-doll-linked-cia-487952>. WEB 7 January, 2016.

Wheatley, Spencer, Thomas Maillart, and Didier Sornette. "The Extreme Risk of Personal Data Breaches and the Erosion of Privacy." *The European Physical Journal* B, 89.7 (2016). <http://arxiv.org/pdf/1505.07684.pdf>. WEB 28 March, 2016.

Whittaker, Zack. "Wikileaks Uncovers TrapWire Surveillance." *ZDNet.com*, August 14, 2012. <http://www.zdnet.com/wikileaks-uncovers-trapwire-surveillance-faq-7000002513/>. WEB 2 October, 2015.

Windrem, Robert and Mike Brunker. "No LOL Matter: FBI Trolls Social Media for Would-Be Jihadis." *NBC News*, October 20, 2014. <http://www.nbcnews.com/storyline/isis-terror/no-lol-matter-fbi-trolls-social-media-would-be-jihadis-n226841>. WEB 4 March, 2016.

Wingfield, Nick and Katie Bender. "Apple Is Rolling Up Backers in iPhone Privacy Fight against FBI." *New York Times*, March 3, 2016. <http://www.nytimes.com/2016/03/04/technology/apple-support-court-briefs-fbi.html>. WEB 4 March, 2016.

Wingfield, Nick and Mike Isaac. "Apple Letter on iPhone Security Draws Muted Tech Industry Response." *New York Times*, February 18, 2016. <http://www.nytimes.com/2016/02/19/technology/tech-reactions-on-apple-highlight-issues-with-government-requests.html>. WEB 22 February, 2016.

Winkler, Sophie. *Open Records Laws: A State by State Report*. National Association of Counties, Research Division. December, 2010. <http://www.naco.org/sites/default/files/documents/Open%20Records%20Laws%20A%20State%20by%20State%20Report.pdf> WEB 12 April, 2016.

Woods, Ben. "What's the True Value of Your Personal Data? Meet the People Who Want to Help You Sell It." *The Next Web*, September 17, 2013. <http://thenextweb.com/insider/2013/09/17/whats-the-true-value-of-your-personal-data-meet-the-people-who-want-to-help-you-sell-it/>. WEB 22 January, 2016.

Zetter, Kim. "Judge Enforces Spy Orders Despite Ruling Them Unconstitutional." *Wired.com*, January 23, 2014. <http://www.wired.com/2014/01/judge-nsl/>. WEB 25 January, 2014.

ABOUT THE AUTHOR

Photo and photo illustration by Scott Cavanah

Edward Lee Lamoureux received a B.A. from CSU, Long Beach (Speech, 1975), M.A. from WSU, Pullman (Speech Communication, 1980), and Ph.D. from the U. of Oregon, Eugene (Rhetoric and Communication, 1985). He arrived at Bradley University (in the Department of Communication) in 1985. He is now also in the Department of Interactive Media.

Ed teaches IM 450: Intellectual Property Law and New Media; IM 355: Theories, Concepts, and Practices in Interactive Media; and IM 450/CFA 314: Privacy in the Connected U.S. Ed also teaches COM 303: Rhetorical Perspectives in Organizational Communication, CFA 201: The Entrepreneurial Mindset in Communications and Fine Arts, and CIV 101: Western Civilization.

His research experiences and interests include intellectual property law, privacy, ethnography/field research, rhetoric, religious communication, conversation, teaching and learning in virtual worlds, and non-linear writing. His publications include *Intellectual Property Law and Interactive Media: Free for a Fee*, 2nd ed. and *Case Analyses for Intellectual Property in New Media* (with Steve Baron and Claire Stewart, for Peter Lang Publishers), numerous journal articles and book reviews (primarily about rhetoric and communication). His first non-academic book project is a work of creative non-fiction, *Threads*.

Ed's creative production includes audio production and web work as well as communication training via digital embellishments. Lamoureux was on the committee that established the Multimedia Program at Bradley, serving as Interim Director/Director of the Program for three years. Lamoureux was the editor of the *Journal of Communication and Religion* (Religious Communication Association) for two consecutive three-year terms, 1998–2003, and now serves on its editorial board.

A member of the Bradley University President's Committee for Entrepreneurship and Innovation, founders of the Turner School of Entrepreneurship and Innovation; Ed is the CFA liaison to the School. He taught the first online course at Bradley (by more than two years) and taught the first Bradley course(s) in virtual worlds (as Professor Beliveau in Second Life). He is a singer/songwriter who enjoys golf and follows the Angels, Lakers, Rams, and Kings. Ed is married (since 1981) to Cheryl; they have four adult off spring: Alexander, Samantha, Kate, and Nicole.

ABOUT THE FRONT COVER

(front cover design concept, Lamoureux)

The Latin inscription at the bottom of the Great Seal (found on the back of a $1 bill, among other places) reads "Novus Ordo Seclorum" and translates to "A New Order of the Ages." The two images around the author's name are eyeshades taken from a watercolor and graphite on paper painting, c. 1936, by H. Langden Brown. The painting is in the public domain.

INDEX

A

abortion, 19
abuse, data, 133–37
access, to data, 67–77
accessibility, 183–84
AccuData, 103
Achenbach, Joel, xv
actions, individual, 195
Acxiom, xxviii, xxxiv, 130
advertising, 31, 121–24, 136, 138
    attitudes toward, xxvi, 142
    business models based on, 141, 176–77
    and inaccurate data, 130–31
    and infrastructure for surveillance, 175–76
    market for, 25
    targeted, 31
affinity analysis, 120
agencies, federal, 150–54. *See also* FCC; FTC
aggregation, of data, 42–43, 49–50, 104
agriculture, 117
algorithm, 39, 120, 135

Alinsky, Saul, 194
Amazon, 93, 119–20
AMC Networks Inc., 11–12
American Civil Liberties Union (ACLU), 16
Amway Global v. Woodward, 77
analytics, 114–18, 129, 200
Angwin, Julia, xxviii–xxix, xxxv, 68–71, 76, 129–130, 194
anonymity, 42–43, 49–50, 126–27, 184–85
Anthem, 2
AOL, 89–90
Apple, 88–90
    App Store, 6, 49
    cooperation with government, 99
    end-to-end encryption, 85
    resistance to government intrusion, 175
    response to data requests, 89, 92–93, 98, 173, 176, 180
apps
    app store, 6, 49
    data collection by, 6
    recommendations for, 171–72

Arias, Myrna, 17
Arias v. Intermex Wire Transfer, 17
ars technica, 10, 47
Article III, 23
Ashley Madison, 3
assembly, freedom of, xiii, 8, 12, 57
association rule learning, 120
AT&T, 176, 180
audience segmentation, 133
Auletta, Ken, 38
Authentify, 7
autonomy, 29–31, 33, 44

B

Baltimore Sun, 59, 89
Bamford, James, 106
Bankston, Kevin, 56, 87, 94–95
Bazzell, Michael, xxx–xxxi, xxxv, 68–71, 76, 194
beacons, 126–27
Benkler, Yochai, 36–37
Berners-Lee, Tim, 145, 177
Beware, 21–22
Binney, William, 180
biometric information technologies, 20
Blizzard Entertainment, 50
Bloomberg, 105
BlueKai, xxxiv
Box, 93
Brandeis, Louis, 13
Branwyn, Gareth, 194
breaches, data, 2–3, 5, 7–8, 52, 116, 137–39
Brill, Julie, 55, 156
Brin, Sergey, 177
"Broadband Consumer Privacy Proposal Fact Sheet" (FCC), 73
Brown, Jerry, 170
browsers, 68, 196
browsewraps. *See* wrap contracts
Bureau of Industry and Security (BIS), 95
business interests. *See* commercial entities; private enterprise
business models, 141–42, 153–54
businesses, local, 178
Butler, Alan, 15

C

cable industry, 150–51
caches, 196
Calamite, Ana, 131
California, 170–72
Calo, Ryan, 108
capitalism, 30, 173
carelessness, consumer, 30–31
Carr, Nicholas, xxxi–xxxii
cars, smart, 17–18
casinos, 184
cell phones. *See* mobile phones
censorship, 177
Center for Civic Media, 85
Centre for Economics and Business Research (Cebr), 25
Cerf, Vint, 145
Chertoff, Michael, 91
ChoicePoint, 104
Christou v. Beatport, LLC, 77
citizens, responsibility for privacy problems, 142
Citron, Danielle, 135–36
civil liberties, 8, 148–49, 161
civil rights, 107–8
class, social, 134
click-throughs. *See* wrap contracts
clickwraps. *See* wrap contracts
cloud-based computing, and breaches, 139
CNET, 150
CNN.com, xxxv, 125–28
Cole, Michael, 19
Cole v. Gene by Gene, Ltd., 19
Commerce Control List (CCL), 94
commercial entities
    cooperation with government, 97–98, 176, 180
    credibility problems, 175–76
    and data requests from government, 88–90, 173–75

privileging of, xxxvi–xxxvii, 44, 67
recommendations for anonymity of data, 184–85
recommendations for defaults, 182–83
recommendations for executive leadership, 176–79
recommendations for industry leadership, 173–76
recommendations for offering multiple versions, 183–84
recommendations for privacy policies/ToS, 181–82
recommendations for worker behaviors, 179–181
recommendations on selling data, 184
responsibility for privacy problems, 142
sharing of data with government, 86–99
and state action, 171–73
Communications Decency Act, 157
Congress, recommendations for legislation by, 154–59. *See also* laws, national
Connected Experience Lab, 115
connectivity, 38
consent, tailored, 166
Consolidated Edison, 68
Constitution, U.S., xxiii–xxiv, 150
    Article III, 23
    Fifth Amendment, 10, 17–18, 57–58
    First Amendment, 8–13, 57, 174
    Fourteenth Amendment, 18–22
    Fourth Amendment, 15–16, 57–58, 153
    ineffectiveness in protecting against surveillance, 51
    and protection of privacy, 8–23, 57–59
    sensitivity to privacy protections in, 164, 167
    Third Amendment, 13–15
Consumer Data Privacy in a Networked World, xxii, 146–49
Consumer Privacy Bill of Rights, xxii, 147
Consumer Watchdog, 77
consumers
    carelessness, 30–31
    cost of data breaches to, 7–8

recommendations for, 191–201
responsibility for privacy problems, 142
content, 142–43, 192
contextual integrity, 34–36
contracts, xxxvi, 44. *See also* privacy policies; terms of service; wrap contracts
control
    of communication system, 36–37
    of health data, 52
    loss of, 2–8, 33
Cook, Tim, 98, 176
cookies, 24, 46, 125–27, 196
Copyright Act (DMCA), 157, 173
CoreLogic, 103
corporate entities. *See* commercial entities; private enterprise
correction, of data, 74, 76
coupons, 134
courts, 158
    applying old laws to new media, 164–65, 167
    in commercial area, 164
    and privileging of corporations, 44
    recommendations for, 163–67
    and ToS/privacy policies, 165–67
    undermining of protections by, 57–59
Cox, Howard, 14
Crain, Matthew, 108, 121, 134
credit cards, xxiv, 195
credit reporting agencies, 154–56, 195
Cronon, William, 102
Crump, Catherine, 10
cryptography, 86–88. *See also* encryption
customer relationship management (CRM), 122
cyber criminals, 139

D

data
    individualized, 24–28
    providers of, responsibility of, 109–11
    re-anonymizing, 185
    re-identifying, 188

selling of, 184
trading for services, 83–85
uses of, 84
data, aggregated, 24
data, de-identified, 125–27
data analysis, real-time, 117
data breaches, 2–3, 5, 7–8, 52, 116, 137–39
data brokers, 70, 156
data collection
   associated with law enforcement/intelligence, 173
   defenses of, 134
   habituation to, 143
   invisibility of, xxvii
   lack of knowledge of details, xxix
   public opinion on, 81–83
   reports about, recommendations for, 156
   and secondary uses, 97
   secrecy of, 8, 42–63
   support for, 31
   See also tracking
data management, xxiv, xxvii–xxviii
data marketers, 70
data marketplace
   companies in, xxxiv
   government/law enforcement utilization of, xxxv
   lack of constraints on, xxv
   secrecy of, 123
data points, treating people as, 133
data requests, 88–90, 173–76, 180, 190–91
data scoring, 71
data streams, mixing of private and public, xxvi–xxvii, xxxvii, 18, 51, 73, 104–9, 148
data wholesalers, 103
Datacoup, 26
Datalogix, 130
Decentralized Web Summit, 145
decision making, rapid, 117
DeepFace, 21
default settings, 39, 182–83
Defense, Department of, 104
democracy
   and autonomy, 29–30
   and capitalism, 30
   and privacy, xiii, xv
   protecting, 159, 200–201
   and surveillance, xiii, xiv, xv
   threats to, 30–31, 142–45, 150, 161, 163, 173
Diffie, Whitfield, 92
Digital Advertising Alliance, 122
digital market manipulation, 108–9
The Digital Person (Solove), 51, 104–5
digital rights management (DRM), and encryption, 87
directories, 71
disadvantaged populations, 107–8
Disconnect, xxxv
discovery, in law, 101–2
discovery, of data, 65, 67–77, 157
discrimination, 107–8, 134–37
DMCA (Copyright Act), 157, 173
DNA tests, 19
Doe v. Southeastern Pennsylvania Transportation Authority (SEPTA), 58
DoubleClick, 46, 121–22
Dougherty, Dale, 24
Dragnet Nation (Angwin), xxix, 129, 194
Drake, Thomas, 59, 180
Dropbox, 90
due process, 18–22, 136
Duggan, Maeve, 82
Dyre approach, 139

E

economy, xxxiii–xxxiv, 148, 158
e-discovery, 101–2
education, 189–191
Eisler, Barry, 99
Electronic Freedom Foundation, 68
Electronic Frontier Foundation, 12, 194
Electronic Privacy Information Center (EPIC), 9, 194
email
   as public record, 72, 102
   recommendations for, 198

employees
- control of social media accounts, 77–79
- privacy of social media accounts, 106
- recommendations for, 179–181

employers, 137
encryption, 85–96, 174–77, 198
Epsilom, xxxv
Equal Employment Opportunity Commission (EEOC), 78
Equifax, 130, 156
errors, in data, 129–133
EU (European Union), xix–xxi, 67, 74–76, 156–57
Eubanks, Virginia, 136
EULAs, 47. *See also* wrap contracts
European Union (EU), xix–xxi, 67, 74–76, 156–57
Evernote, 90
Excellus BlueCross Blue Shield, 3
eXelate, xxxiv
Experian, 3, 130, 138, 156
export restrictions, and encryption, 88, 93–96
Exposed (Harcourt), 194

F

FAA (Federal Aviation Administration), 161
Facebook, xxxv, 74–76, 88–90, 93, 98, 110, 119, 124, 185
- advertising on, 176–77
- DeepFace, 21
- and facial recognition technologies, 20–21
- provision of Internet access, 85
- purchase of WhatsApp, 175
- and right to be forgotten, 70
- tracking by, 11
- and value of data, 26–27
- *See also* social media

facial recognition technologies, 19–21
Fair and Accurate Credit Transactions Act (FACTA), 65–67, 154–56
Fair Credit Reporting Act (FCRA), 23, 65–67, 154–56

Fair Information Practices (FIPs), xviii–xix, xxi, xxii
- and education, 190
- FCRA's similarities to, 155
- and financial data, 65–67
- FIP 1, 41–63
- FIP 2, 65–79, 103
- FIP 3, 81–111
- FIP 4, 65–79
- FIP 5, 113–39
- following, 178–79
- importance of, 118, 145–46
- as indicators of norms, 35
- in medical releases, 53–54
- secrecy's challenges to, 50
- and states, 168–171

Family Educational Rights and Privacy Act (FERPA), 190
FBI
- and facial recognition technologies, 20
- investigations into 9/11, 162

FCC (Federal Communications Commission), 73, 150–52, 163
FCRA (Fair Credit Reporting Act), 23, 65–67, 154–56
FDA (Food and Drug Administration), 51
Federal Aviation Administration (FAA), 161
Federal Communications Commission (FCC), 73, 150–52, 163
Federal Trade Commission (FTC), xxi–xxii, 117–18, 147, 152–54, 156, 163, 184
Fernandes, Teresa, 131
FERPA (Family Educational Rights and Privacy Act), 190
Fifth Amendment, 10, 17–18, 57–58
financial data, 65, 154–56
FIPs (Fair Information Practices). *See* Fair Information Practices
First Amendment, 8–13, 57, 174
FISA court (Foreign Intelligence Surveillance Court), 61–62, 89, 97, 159, 164
FISC/FISA court. *See* FISA court
fitness monitors, 84–85, 188, 198–99
Flaming, Todd, 101
FOIA (Freedom of Information Act), 99

Food and Drug Administration (FDA), 51
Foreign Intelligence Surveillance Court (FISA court), 61–62, 89, 97, 159, 164
forgotten, right to be, 69–70, 76–77, 156–57
Fourteenth Amendment, 18–22
Fourth Amendment, 15–16, 57–58, 153
Freedom of Information Act (FOIA), 99
Fresno, California, 21
Frye, Curtis, 100
FTC (Federal Trade Commission), xxi–xxii, 117–18, 147, 152–54, 156, 163, 184
functionality, 124–29

G

Gangadharan, Seeta Peña, 136
Gartner, 130
Gatto, Mike, 13–14
GDCDPA (Government Data Collection and Dissemination Practices Act), 168–71
GE, 115
Gellman, Barton, 89
Giannandrea, John, 24
The Glass Cage (Carr), xxxi–xxxii
golden mean, 30
Goodin, Dan, 106
Google, 74–75, 88–90, 93, 98, 119, 124, 131–32
    advertising on, 176–77
    DoubleClick, 122
    and facial recognition technologies, 20–21
    recommendation system, 120–21
    and regulatory actions by EU, 74
    and right to be forgotten, 76–77, 156–57
    role in data collection/management, 69
    terms of service, 44–46
    Webpage Screenshot, 47
Google Analytics, 46
Google Spain v. Agencia Española de Protección de Datos (AEPD) and Mario Costeja González, 75–76
government
    commercial entities' sharing of data with, 51, 86–99
    constraint from predatory practices, 168–171
    data collection by, 59–60
    data requests by, 88–91, 173–76, 180, 190–91
    lack of federal standards in privacy law, 167–68
    legality of data collection programs, 60–61
    recommendations for executive leadership, 146–150
    recommendations for federal agencies, 150–54
    recommendations for intelligence community, 159–163
    recommendations for judges/courts, 163–67
    recommendations for law enforcement, 159–163
    recommendations for legislation, 154–59
    recommendations for states, 168–173
    responsibility for privacy problems, 142
    sharing of data with commercial entities, 99–104
    transparency of, 51, 61
    utilization of private data marketplace, xxxv, 105–7
    See also intelligence community; law enforcement; national security; National Security Agency
Government Data Collection and Dissemination Practices Act (GDCDPA), 168–171
Greenwald, Glenn, 105
Guardian, 89
guidelines for handling electronic records and databases, xvii. *See also* Fair Information Practices
Gunnarsson, Helen W., 101

H

Hacked (website), 2
hacking, 2–4. *See also* data breaches

Hamilton, Greg, 129, 131
Handshake, 26
Harcourt, Bernard E., 132, 194
Hardy, Quentin, 145
Health, Education, and Welfare, Department of (HEW), xvii, 52
health information, 19, 43, 51–55, 186–87
   consumer-generated, 55, 85, 188
   control of, 52
   data breaches, 4
   and fitness monitors, 85, 188, 198–99
   leaking of, 136
   patient portals, 188
   uses of, 116–17, 153
Health Insurance Portability and Accountability Act (HIPAA), 52, 55, 85, 136, 186, 188, 190, 199
healthcare industries, 115–17. See also nonprofit entities
Hello Barbie doll, 14
Henderson, Wade, 108
Herbert, Gary, 170
HEW (Health, Education, and Welfare), xvii, 52
Heyman, David, 98
Hiding from the Internet (Bazzell), xxx–xxxi, 194
HIPAA (Health Insurance Portability and Accountability Act), 52, 55, 85, 136, 186, 188, 190, 199
home, data collection in, 13–15
Hoofnagle, Chris, 11
Hook, Nigel, 117
Howe, Scott, xxviii
human rights, 63

I

IBM, 116
ICANN (Internet Corporation for Assigned Names and Numbers), 4
identity theft, 7, 138
Identity Theft Resource Center (ITRC), 138
Illston, Susan, 12

inaccuracy, of data, 129–133
information, and need for profit, 32–33
information, freedom of, 157
information, personal, xxix–xxxi. See also data
infrastructure, for surveillance, 175–76, 192
innovation, 153–54, 165
Instagram, 49, 185
integrity, contextual, 34–36
intellectual property, xxxvi
intelligence community
   adherence to legal structures, 159–61
   FISA court, 61–62, 89, 97, 159, 164
   and judges, 164
   programs, 58–59
   recommendations for, 159–63
   surveillance of innocent people, 161–62
   Total Information Awareness, 58–59, 104–5
   total surveillance power of, 144
   use of new technologies, 160, 162–63
   use of third-party data, 106–7
   See also government; national security
interface, 39
Internet
   accessing, 85, 196, 197
   commercialization of, 38
   excluded from FCC oversight, 151–52
   and increase in tracking, xxiv
   protection of by courts, 165
   value of, 123
Internet Corporation for Assigned Names and Numbers (ICANN), 4
Internet industries. *See* commercial entities; private enterprise
Internet of Things (IoT), 14, 84, 114–15
   and likelihood of breaches, 139
   potential for harms to consumers, 118
   and privacy, 117–18
   recommendations for, 175, 198–99
Internet Service Providers (ISPs), 11, 72–73
invasiveness, commercial, 30–31
IoT (Internet of Things). *See* Internet of Things
IP address, masking, 196

Ireland, 174
ISPs (Internet Service Providers), 11, 72–73

J

Jamming the Media (Branwyn), 194
Jenkins, Holman, 131
Jouvenal, Justin, 21
judges, 163–67. See also courts

K

Kahle, Brewster, 145
Kehl, Danielle, 87, 94–95
Kim, Nancy S., 43–44, 166
Kirkpatrick, David, 38
Klein, Mark, 180
Kromtech, 5

L

Landreth, William, 2
Larkin, Helma, 176
law enforcement, xxxvii
    adherence to legal structures, 159–61, 165
    constraint from predatory practices, 168–171
    dependence upon federal assistance, 168
    limits on data collection by, 170
    recommendations for, 159–63
    surveillance of innocent people, 161–62
    use of big data, 21
    use of new technologies, 160, 162–63
    use of social media, 106–7
    utilization of private data marketplace, xxxv
laws
    lag behind technology, 167, 172
    for old media, 164–65, 167
    Privacy Act of 1974, xix
    and protection to citizens, xxxvi–xxxvii

laws, local, 168
laws, national
    lack of, 158
    risks of, 167–68
laws, state, 168, 170
Lazarus, David, 67
leadership, benevolent, 30–31
leapfrogging, 139
legal protections, 43. See also Constitution, U.S.
legislation
    and intellectual property, xxxvi
    need for, 118
    recommendations for, 154–59
    See also laws
Lessig, Lawrence, xxxix, 29, 37
Levie, Aaron, 93
Levy, Steven, 87
LexisNexis, xxxv, 166
Libert, Tim, 136
libraries, 33, 190–91
license plate readers, 10–11, 170
LifeLock, 5
LinkedIn, 90, 137
litigation, and abusive data management practices, 23
location data, 9–11, 84, 170, 185
low-income citizens, 107–8
loyalty cards, 195

M

MacKeeper, 5
Madden, Mary, 83
Maillart, Thomas, 139
Maras, Elliot, 2
market manipulation, 108–9
marketing, 49, 71, 121–24, 136, 138
    attitudes toward, xxvi, 142
    business models based on, 141
    defenses of, 134
    inaccuracies, 129, 131, 133
    and infrastructure for surveillance, 175–76
    and misuse of data, 133–37

INDEX

Marketing Letters, 114
Marshall, Jack, 123
McCarthy, John, 92
McChesney, Robert W., 30, 62, 106, 124
McLuhan, Marshall, xxxi, 29, 31–32, 40, 142–43, 192
McPhail, Michael J., 2
medical information. *See* health information
medical releases, 53–54
medium, 142–44, 192
"Meet the Digital Dissenters" (Achenbach), xv
Merge Healthcare, 116
(meta)data, 39
metadata, 15–16
Microsoft, 48, 88–90, 93, 98, 174–75
Miller, Vincent, xxxiii, 133
minorities, 107–8
mobile phone providers, role in data collection, 73
mobile phones
    accessing Internet on, 196–97
    and increase in tracking, xxiv
    protection of conversations, 165
    recommendations for, 197–98
    searches of records, 15–16
    stingray devices, 16, 170
    use of data, 21
Montulli, Lou, 24
morality, evaluating, 35
MySpace, 153

N

Naeimi, Helia, 138
Narayanan, Arvind, 50
national security, xxxvii, 149, 158–59.
    See also government; terrorism
National Security Agency (NSA), xxv, 56, 59, 62, 86–87, 91, 96–98, 105–6, 144, 161–62, 180. *See also* Snowden, Edward
National Security Letter (NSL), 12

Negroponte, Nicholas P., 40
Netflix, 119–20
Network, 25
Neuburger, Jeffrey, 20
New Hampshire, 170
new media infrastructure, 30–34
New York Times, 50, 145, 175
9/11, xxv, xxxvii, 160–63, 190.
    See also terrorism
Nissenbaum, Helen, 34, 114
Nixon, Richard, 143–44
nonprofit entities, 129, 185–191
norms, and judging propriety of privacy/surveillance, 34–35
NSA (National Security Agency). *See* National Security Agency
NSL (National Security Letter), 12

O

Oakland, California, 10–11
Obama, Barack, 148–50, 163
Obama administration, 146, 168
O'Dell, Jolie, 193
Office of Personnel Management, 3
Olmstead v. United States, 13
open records, 171
Open Records Law (Winkler), 171
Opsahl, Kurt, 87
Oregon, 170
O'Reilly, Tim, 24, 84
Ornstein, Charles, 55

P

pacemakers, 117
Page, Lawrence, 177
PalTalk, 89
Parker, Donn, 2
partners, 125, 128
Patriot Act, 88–90, 97, 144, 159, 161, 190
PaymentsMD, 153
Peacock, Sylvia, 135

people, the
    recommendations for, 191–201
    responsibility for privacy problems, 142
    See also consumers
Peppers, Don, 122
Personal.com, 26
Petersen, J. Chapman, 169
Peterson, Andrea, 56
Pew Research Center, 81
PhoneDog LLC v. Kravitz, 77
Pinterest, 185
place, cyberspace as, 8
Poitras, Laura, 89, 105
political action, recommendations for, 199–200
politics, and protection to citizens, xxxvi–xxxvii
Ponemon Institute, 7
Postal Service, United States, 103
Postman, Neil, 32–33, 34, 40, 162
poverty, 85
powerboat racing, 117
predictive analysis, 132–33
Premera Blue Cross, 3
president, United States, 143–44, 150. *See also* Obama, Barack; Obama administration
price discrimination, 108
pricing strategies, 134
prior restraint, 8, 12–13, 174
PRISM, 89
privacy
    aspects of, xvii–xviii
    attitudes toward, 110, 123
    and autonomy, 29–30
    loss of, lack of concern for, xxiii
    and norms, 34–35
    right to, 18–22, 153
    value of, xxxviii–xxxix
Privacy Act of 1974, xix
"Privacy and Information Sharing" (Pew Research Center), 81
Privacy on the Go, 171
privacy policies, xxix, xxxi, 49–50, 119, 186–87
    CNN.com's, 125–28
    deceptive, 153
    and FTC, 152–53
    hiding of data collection/management, 47
    and judges/courts, 165–67
    presented on paper, 187
    recommendations for, 154, 181–82
    Spotify's, 47–48
    Uber's, 9–10, 153
    and usability and functionality lie, 125
Privacy Rights Clearinghouse, 194
private enterprise
    data collection and management by, 42
    and lack of constraints on data marketplace, xxv
    relationship with government surveillance, 51, 62–63
    See also commercial entities
problems, solved by new technology, 32–33, 162
profiling, 135
profit, capitalism's need for, 30
ProPublica, 55
protocol, 39
Proton Mail, 198
public entities, 42. *See also* government
public records, 71–73, 99–104, 170–71, 196

Q

Qriously, 123
Quibell, Marc, 109

R

Rainie, Lee, 82
Rall, Ted, 98
Ramirez, Edith, 152–54
Rapleaf, xxxiv
re-anonymizing data, 185
recommendation systems, online, 118–121
recommendations
    for apps, 171–72

for commercial entities, 173–91
for Congressional legislation, 154–59
for email, 198
for executive leadership in government, 146–50
for federal agencies, 150–54
for government, 146–73
for industry leadership, 173–76
for intelligence/law enforcement, 159–163
for IoT, 175, 198–99
for judges/courts, 163–67
for leadership of commercial entities, 176–79
for legislative action on intelligence activities, 158–59
for mobile phones, 197–98
for nonprofit entities, 185–91
for the people, 191–201
for social media, 196–97
for states, 168–73
for worker behaviors, 179–81
Records, Computers, and the Rights of Citizens (HEW), xvii–xviii. *See also* Fair Information Practices
recruitment, by employers, 137
Reed Elsevier, xxxv
re-identifying data, 188
reliability, of data, 113–14, 129–33
religion, freedom of, xiii
removal, of data, 157
repair, of data, 65
resources, 194, 196, 198
responsibility, for existence of data, 109–11
revenue, from selling data, 184
RFID technology, 117
Richards, Neil, 135
right to be forgotten, 69–70, 76–77, 156–57
Rigley, Michael, 25
Riley v. California, 15
Risen, James, 105
Robins, Thomas, 23
Robinson, Stephen Cory, 134
Roe v. Wade, 19
Rogers, Martha, 122

Rotenberg, Marc, 15
Rules for Radicals (Alinsky), 194
Ryan, Johnny, 38

S

Safire, William, 58–59
San Bernardino County, California, 16
Schmidt, Eric, 131–32
schools, 185, 189–90. *See also* nonprofit entities
Schrems, Max, 75
searches and seizures, 15–16
secondary uses of data, 81–83, 97, 101–2
secrecy
    challenges to FIPs, 50
    and data anonymity via aggregation, 49–50
    of data collection, 8, 42–63
    of data marketplace, 123
    and government surveillance, 30, 56–57, 61–62
    in wrap contracts, 42–49
secret, defined, 41
SeisInt, 104
self-incrimination, 11, 17–18, 57
"Self-Regulatory Principles for Online Behavioral Advertising" (FTC), xxi–xxii
sentiment analysis, 132–33
set-top box manufacturers, 150–51
Shmatikov, Vitaly, 50
shrink wraps. *See* wrap contracts
Singer, Natasha, 68, 76
Skype, 89
Snowden, Edward, xiv, xxvi, 55–56, 73, 82, 89, 93, 96–98, 104–5, 148, 180. *See also* National Security Agency
social media
    benefits of connecting customers to, 186
    citizen participation in, xxv–xxvi
    finding information on, 69–70
    law enforcement's use of, 106–7
    and nonprofit entities, 185–86, 188–89

opting out of, 193
ownership of accounts, 77–79
and political action, 199
privacy of accounts, 106
recommendations for, 196–97
social media companies
  data given to, 109–10
  and First Amendment, 12–13
  and government surveillance, 12
  See also Facebook; Twitter
social networks, professionalized, 137
Soghoian, Chris, 153
soldiers, quartering of, 13
Solove, Daniel J., 8, 50–51, 58, 104–5
Sornette, Didier, 139
sorting, 135
speech, anonymous, 57
speech, freedom of, xiii, 8–9, 12, 57, 157, 174
Spokeo v. Robins, 22–23, 50
Spotify, 47–48
spying. *See* national security; surveillance states
  and commercial entities, 171–73
  legislation by, 170
  recommendations for, 168–173
  role in data collection, 170–71
stingray devices, 16, 170
subpoenas, 158
Supreme Court, U.S., 22–23, 50
surveillance, xxvi, 55–57
  authority for, 149–50, 158, 161
  in history, xxiii–xxiv
  legislative efforts to constrain, xxiv
  and norms, 34–35
  power of, 144
  support for, 31
surveillance agencies, xxv. *See also* National Security Agency

T

Tanner, Adam, 124, 146, 184, 194
Target, 6–7

targets vs. waste, 133–37
tech companies. *See* commercial entities; social media companies
technology, xxxii–xxxiii
telecommunication companies, 12–13, 150–51
10 Minute Mail, 198
Tene, Omer, 147
terms of service (ToS), xxix, xxxi, 128, 186–87
  Google's, 44–46
  hiding of data abuse, 54
  hiding of data collection/management, 47
  and judges/courts, 165–67
  presented on paper, 187
  recommendations for, 154, 181–82
  See also wrap contracts
terror, war on, 158, 163
terrorism, xxv, 56, 144, 149, 160–63, 173, 175, 190. *See also* national security
text messaging, 197
Third Amendment, 13–15
third parties, 125, 128
third-party data, 104, 106
third-party doctrine, 15
threat scoring system, 21–22
T-Mobile, 3
ToS (terms of service). *See* terms of service; wrap contracts
Total Information Awareness (TIA), 58–59, 104–5
totalitarianism, 159
ToyTalk, 14
tracking, 196
  amount of, xxviii–xxix
  blocking, xxxv, 68–69
  difficulty in managing, xxix–xxxi
  and Fifth Amendment, 17
  increase in, xxiv, 84
  of offline activities, xxviii
tragedy of the commons, 135
Trailblazer, 59
transparency, 42, 51, 61

TransUnion, 130, 156
trust, of commercial entities, 173, 182
Turow, Joseph, xxxiv, 38, 71, 107, 123, 134
Twitter, 12–13, 70, 98, 174. *See also* social media; social media companies

U

Uber, 9–10, 153
United Nations, 91
United States v. Wurie, 15
UnityPoint-Methodist, 186–88
unreliability of data, 129–133
USA Freedom Act, 97
usability, 124–29
users, 30
    carelessness, 31
    recommendations for, 191–201
    responsibility for privacy problems, 142
Utah, 170

V

value equation, and personal data, 23–28
Van Dijck, José, 38–39
Venture Beat, 193
Verizon Wireless, 68, 89, 151
Vickery, Chris, 5
vineyards, 117
Virginia, 168–71
voice recognition software, 132
VTech, 3

W

Walker, Scott, 72, 102
Wall Street Journal, xxviii, 108, 123, 131
war on terror, 158, 163
Warner, Bernhard, 106
warrants, 16, 158
Warren, Samuel D., 13
Washington Post, xv, 21, 89

waste vs. targets, 133–37
Watson Health, 116
The Wealth of Networks (Benkler), 36
wearables, 84–85, 116–17, 188, 198–99
Web 2.0, 24, 39, 83–85, 141
web beacons, 126–27
weblining, 134
Webpage Screenshot, 47
Wertheimer, Michael, 91
Whalen, Andrew, 14
What Stays in Vegas (Tanner), 194
WhatsApp, 175
Wheatley, Spencer, 139
"Where 2.0" conference, 84
White House white paper, 146–49
white paper, 146–49
Wiebe, J. Kirk, 180
WikiLeaks, 3–4
Wilson, Andi, 87, 94–95
Windows 10, data collection by, 48
Wired.com, 47
Wisconsin, 72, 102
wrap contracts, xxxvi, 42–49, 53–54, 73, 128, 152–53, 166, 181–82
Wrap Contracts (Kim), 43–44
Wyden, Ron, 158–59

Y

Yahoo, 89–90, 98
YouTube, 89

Z

Zogby Analytics, 123
Zuckerberg, Mark, 177
Zuckerman, Ethan, 85

General Editor: **Steve Jones**

Digital Formations is the best source for critical, well-written books about digital technologies and modern life. Books in the series break new ground by emphasizing multiple methodological and theoretical approaches to deeply probe the formation and reformation of lived experience as it is refracted through digital interaction. Each volume in **Digital Formations** pushes forward our understanding of the intersections, and corresponding implications, between digital technologies and everyday life. The series examines broad issues in realms such as digital culture, electronic commerce, law, politics and governance, gender, the Internet, race, art, health and medicine, and education. The series emphasizes critical studies in the context of emergent and existing digital technologies.

Other recent titles include:

Felicia Wu Song
  Virtual Communities: Bowling Alone, Online Together

Edited by Sharon Kleinman
  The Culture of Efficiency: Technology in Everyday Life

Edward Lee Lamoureux, Steven L. Baron, & Claire Stewart
  Intellectual Property Law and Interactive Media: Free for a Fee

Edited by Adrienne Russell & Nabil Echchaibi
  International Blogging: Identity, Politics and Networked Publics

Edited by Don Heider
  Living Virtually: Researching New Worlds

Edited by Judith Burnett, Peter Senker & Kathy Walker
  The Myths of Technology: Innovation and Inequality

Edited by Knut Lundby
  Digital Storytelling, Mediatized Stories: Self-representations in New Media

Theresa M. Senft
  Camgirls: Celebrity and Community in the Age of Social Networks

Edited by Chris Paterson & David Domingo
  Making Online News: The Ethnography of New Media Production

To order other books in this series please contact our Customer Service Department:
  (800) 770-LANG (within the US)
  (212) 647-7706 (outside the US)
  (212) 647-7707 FAX

To find out more about the series or browse a full list of titles, please visit our website:
  WWW.PETERLANG.COM